Evolutionary Theory and the Creation Controversy

Olivier Rieppel

Evolutionary Theory and the Creation Controversy

🐎 Springer

Dr. Olivier Rieppel
Department of Geology
The Field Museum
1400 S. Lake Shore Drive
Chicago, Illinois 60605-2496
USA
orieppel@fieldmuseum.org

ISBN 978-3-642-14895-8   e-ISBN 978-3-642-14896-5
DOI 10.1007/978-3-642-14896-5
Springer Heidelberg Dordrecht London New York

Cover design: deblik, Berlin

Printed on acid-free paper   B 2880859

Springer is part of Springer Science+Business Media (www.springer.com)

Preface

This book is about the evolution of evolutionary theory against the background of Creationism and Intelligent Design. It looks at evolutionary theory from a historical perspective in an attempt to clarify some very basic issues surrounding evolutionary theory, yet issues that are rooted in the metatheoretical background of this theory. What is the metatheoretical background of a scientific theory?

Evolutionary theory has become a highly complex, multifaceted body of thought. Subdisciplines may sometimes appear partially contradictory to the layperson; some require special training for a proper understanding. And yet, evolutionary theory has become a predominant worldview in modern times. Evolutionary theory primarily addresses the problem of the origin of the diversity of plant and animal species that we observe today. In more recent times, however, evolutionary theory has gained currency far beyond its original confines. Attempts to understand the origin of the earth, indeed of the solar system, are now cast in an evolutionary context. And so are attempts to understand the origin and historical development of human culture, civilization, and language; the origin of the powers of human cognition; and even the origin of moral and ethical values guiding and constraining everyday life in human society. Engineering uses computer software to simulate evolutionary processes such as (natural) selection in the attempt to optimize the design of complex mechanical systems such as aircraft. Natural selection theory is put to use in the development of vaccines.

Charles Darwin explained the origin of new species through a process of random variation and natural selection. Those variations would be favored and hence propagated, which would provide for a slightly better adaptation to the environment, i.e., those that would have a selective advantage in terms of relative reproductive success. As the eminent evolutionary biologist Ernst Mayr never tired to emphasize, this is a statistical issue, an issue of relative numbers of offspring, not an issue of good versus bad, and nothing in between. Stripped to its essentials, the theory of natural selection explains the adaptation of organisms to their particular environment, but it does not necessarily imply progressive evolution to higher levels of complexity or optimization of biomechanical design. The notion of progressive evolution, of perfection of design – notions that Darwin himself entertained at least during the early stages of the development of his theory – is not inherent in the Darwinian theory of natural selection. It is added to that theory as a belief rather than as a theoretical component supported by the study of nature. Progressionism is one metatheoretical burden imposed on evolutionary theory, but it is only one among many. Others are Social Darwinism, the slogan of the "survival of the fittest, " and Lord Alfred Tennyson's metaphor of "nature red in tooth and claw," the idea that historical or sociological changes proceed gradually and in small steps for the benefit of society or the belief that human powers of cognition have been shaped by natural selection and hence are adapted to cognize the world as it really is, at least to a degree assuring the survival of the human race. Moral as well as ethical standards guiding and constraining human social behavior are claimed to have an evolutionary basis – but how could a morally indifferent natural process such as biological evolution give rise to the intellectual insight that the way things "are" may not always be the way things "ought to be," if judged from an ethical point of view?

Small wonder that many were left dissatisfied with evolutionary theory, when in fact they were opposing the metatheoretical baggage imposed on that theory. Opponents of evolutionary theory may have read this metaphysical baggage into evolutionary theory the way they understood it and hence may simply have set up a straw man in order to knock it down. In many cases, however, popular accounts of evolutionary theory are loaded with the same metatheoretical baggage and are justly criticized on that account. Most importantly, however, contemporary evolutionary theory has become so intricate and complex that it is often hard to distinguish between scientific theory and metatheoretical beliefs, even for the informed critic.

This is why I propose to look back on two books published during the middle of the nineteenth century, published a few years before Darwin's *Origin of Species*. The two books deal with an evolutionary theory and its criticism yet in a context that was much easier to understand than it is today. Both books, pro and contra evolution, were written by educated laymen, who presented their arguments in a manner highly transparent to a general readership. However, the same metatheoretical baggage is still with us today: is evolution subject to any lawfulness, is evolution progressive, is adaptation perfect? But the context is much easier: no theory of inheritance had yet been developed, the meaning of fossils was only beginning to become clear, and the mechanisms of embryonic development still remained a black box. Black boxes are suitable containers for projections of the mind, and this is why we find these black boxes filled with beliefs and projections that still concern us today.

The two books I want to discuss are Robert Chambers' *"Vestiges of the Natural History of Creation"* (1844), an early, perhaps even "naïve" (if viewed from a modern perspective), enactment of evolutionary theory, yet a book that provoked stern criticism, as voiced by Hugh Miller in his *"Foot-prints of the Creator"* (1849). I propose to look at these two publications not as mirrors of Victorian culture in nineteenth Century Great Britain, nor as the state of the art of professional science in pre-Darwinian nineteenth century biology and geology, nor do I plan to place their authors into their proper social and political contemporary settings to elucidate their broader motives and concerns as they published these books. Readers interested in these aspects of science and its history should consult treatises such as James A. Secord's *"Victorian Sensation,"*[1] or Martin J.S. Rudwick's two companion volumes *"Bursting the Limits of Time"*[2] and *"Worlds before Adam."*[3] Instead, I propose to screen these two books for conflicts of interest or intellect that are still with us and yet use the less complex context of the nineteenth century debate to bring out the essentials of these conflicts in better contrast than would be possible in the context of modern "population biology" or "phylogenetic analysis." At the same time, this approach will allow us to investigate how Darwin dealt with the issues raised by these two authors in an attempt to preempt criticism of his own theory that might be voiced by the informed public. Note that Chambers' *"Vestiges"* was published in the same year (1844) in which Darwin completed a lengthy essay on his theory of evolution, instructing his wife to have it published in case of his premature death.[4] Naturally, the clash between Chambers and Miller in the mid-nineteenth century played out in the context of a conflict between creationist ideas opposing materialist theories of species transformation – a conflict that has become prominent again in modern times.

In the nineteenth century, as well as before and after, there were two views of God's involvement with the works of nature. The doctrine known as Theism invokes a God who is personally and actively involved in the natural processes. The problem here is that an eternal Entity is supposed to enter time in space, when from a logical point of view, "you cannot make temporal 'events' out of 'eternal objects' without impairment of the eternality of those objects."[5] Deism is the doctrine that God directs the natural course of events through the enactment of natural laws, the so-called secondary causes. There is no conflict studying the natural course of events, while believing that what we discover to be natural laws do in fact lead back to a First Cause. It must be recognized however, that moving the discussion from secondary causes to the First Cause means to *"go meta,"* as philosophers say. It means to move up one level: we move from science to a metascientific level of discourse, i.e., a discourse that extends beyond the reach of the arm of science. The discourse within science is about which natural laws range over which natural processes. At the metascientific level, the discourse is no longer about the relation of natural laws to natural processes. Natural processes do

[1]Secord, J.A., 2000. Victorian Sensation. The Extraordinary Publication, Reception, and Secret of *Vestiges of the Natural History of Creation*. The University of Chicago Press, Chicago.

[2]Rudwick, M.J.S. 2005. Bursting the Limits of Time. The Reconstruction of Geohistory in the Age of Revolution. The University of Chicago Press, Chicago.

[3]Rudwick, M.J.S. 2008. Worlds Before Adam. The Reconstruction of Geohistory in the Age of Reform. The University of Chicago Press, Chicago.

[4]Egerton, F.N. 1970. Refutation and Conjecture: Darwin's Response to Sedwick's Attack on Chambers. Studies in History and Philosophy of Science, 1: 176–183.

[5]Lovejoy, A.O. 1930 (1977). The Revolt Against Dualism. Open Court, Chicago, p. 137.

not figure any more in this upper level discourse. Instead, the discourse is about the natural laws themselves and about how they relate to a First Cause.

Thomas Aquinas (also known as Thomas) famously elaborated on the fact that the human foot and the horse's foot are perfectly adapted – each in its own way – for the purpose that they had been designed for by the Creator. The idea that organisms are built according to a blueprint is Creationist in nature. The organism is compared to a complex yet perfect machine ("clockwork"), built to obey laws of nature or secondary causes, and one cannot tinker with only one or the other part of the machine without interfering with its proper function. Either the machine remains the way it is, or an entirely new clockwork has to be designed ("created").

Intelligent Design is a sophisticated version of Creationism that does not tie God to particular space–time regions. The argument is more of a "First Cause→secondary causes" type. The idea is that natural selection does or can work, but this theory takes us only so far. It is said that there are structures in nature that are simply too complex to be the result of chance variation and contingent natural selection. Therefore, some other forces must be at work in nature that lead back to the First Cause. The issue centrally at stake here is the notion of the "explanatory power" of scientific theories. It is often said that proponents of "Intelligent Design" redefine science. What proponents of "Intelligent Design" in fact do is to gloss over the step where they "*go meta*"; they blur the proper distinction between different levels of discourse.

Olivier Rieppel

Contents

Chapter 1
What Is the Story to be Told?

The famous 19th century German embryologist Karl Ernst von Baer, a contemporary of Darwin and an opponent of Darwin's theory of evolution, is reported to once have aptly characterized the scientific attitude as follows: "If in the course of scientific research I find myself using a balance, and God intervenes by pushing the lighter scale down so that it appears to be the heavier one, I'll have to get up and say – 'please, Sir, you complete the task at hand; under the present circumstances I cannot do it!'."

1.1   The Encyclopedia of Life: Noah's Ark Digitized

"A web page for every species known to live on Planet Earth": this was the vision developed by the renowned Harvard biologist Edward O. Wilson.[1] But with approximately 1.8 million species named and described, and many more yet to be discovered, this would not be an easy project. Each species would have its own web page, and as the species itself, its web page would continuously evolve, "dynamically synthesizing content ranging from historical literature and biological descriptions to stunning images, videos, and distribution maps."[2] This is the mission of the project that has its roots in Edward O. Wilson's proposal: the Encyclopedia of Life. The website of the project[2] invites everybody to help and contribute to the Encyclopedia of Life. This sounds familiar as such a public encyclopedia already exists on the worldwide web: "Wikipedia," the free encyclopedia with entries that are written collaboratively by volunteers around the world. Everybody can log in and enter a new article, or add to or edit an already existing one. If it is possible to motivate systematists, conservation biologists, bird watchers and butterfly collectors to contribute, the Encyclopedia of Life could become a dream come true: an

---

[1]Wilson, E.O. 2003. The encyclopedia of life. Trends in Ecology & Evolution, 18: 77–80.
[2]http://www.eol.org

O. Rieppel, *Evolutionary Theory and the Creation Controversy*,
DOI 10.1007/978-3-642-14896-5_1, © Springer-Verlag Berlin Heidelberg 2011

exhaustive catalogue of life, dubbed "a virtual Ark of Noah" in the media.[3] Would
Noah's Ark, afloat in cyberspace, contain an inventory of Divine Creation? Or a
catalogue of the biodiversity created through the process of the evolution of life on
earth? What is the difference?

Well, it depends on the purpose and goal of the Encyclopedia of Life. If an
illustrated list of every species were all there is to the Encyclopedia of Life, it would
not really matter how species came into being. Whether species originated through
Divine Creation or through descent with modification, they are essential parts of our
natural environment and hence worth every effort to protect them. But to protect
endangered species, one must have deep knowledge about them. An enriched
Encyclopedia of Life thus promises to become a formidable tool for all those active
in nature conservation programs. And yet, like any other dictionary, the Encyclopedia of Life would require some structure and some organization for it to become
an easy-to-use tool for scientists and laypeople alike. Dictionaries are organized in
alphabetical order. So species entries into the Encyclopedia of Life could be also
ordered alphabetically. This was, indeed, the suggestion of one of the leading
initiators of the whole project. One could imagine that Noah released animal
species from his Ark in the alphabetical order of their names, or that reports on
biodiversity assessments of remote rain forest territory list plant and animal species
in alphabetical order. Similarly, one can imagine searching for a species in the
Encyclopedia of Life according to where its name slots into the alphabetical order
of all existing species names.

But then, there is something counterintuitive about such an alphabetical listing
of species names. Children quickly learn the difference between dogs and cats,
lizards and snakes, birds and bats. Visitors to a zoo expect to see eagles in the
birdhouse, tree shrews in the mammal house, and fishes in the aquarium. In nature,
species do not group in the alphabetical order of their names, but instead form
groups marked out by their "nature." The neighbor's cat and a dog that lives further
down the street might be of similar size, and yet the cat seems to be closer to a lion,
whereas the dog shares its characteristics with wolfs. Both birds and bats are warm-
blooded, and yet bats seem to be more closely related to mice than to sparrows.
There is a reason for this, which is spelled out by evolutionary theory: bats and mice
share a common ancestor, which is not shared by birds. With evolutionary theory in
place and well researched for more than 150 years, there opens a possibility to order
plants and animals not only alphabetically but on the basis of kinship. Descent with
modification is the process that governs the growth of the Tree of Life. On this tree,
species nest according to their evolutionary relationships.

The science that groups species in a hierarchically structured natural system according to their evolutionary relationships is called systematics. But to
systematically arrange the Encyclopedia of Life requires expert knowledge.

[3]Radio-Canada http://www.radio-canada.ca/nouvelles/Science-Sante/2007/05/09/001-encyclope-
die-vie.shtml?ref=rss (no implication of 'Intelligent Design' or 'Science Creationism' is intended.
Accessed November 9, 2009).

However, this sets limits to the community of authors competent enough to enter species descriptions into the Encyclopedia of Life, if the latter is to be organized so as to reflect the structure of the Tree of Life. Bird watchers and beetle collectors might be highly competent to enter species descriptions in alphabetical order, or edit the one that already exists, as new information is gathered. But they might be unaware of, perhaps not even interested in, the latest analysis of the evolutionary relationships of the species that they observe and collect. A manual on the proper management of natural history collections, published by the US National Park Service, suggests a variety of arrangements that can be used to organize bird and mammal collections. These may take the form of an arrangement of species according to their evolutionary relationships as specified by a "recognized taxonomic authority,"[4] or an alphabetical arrangement of species within their family or genus. A dispute thus arose amongst the initiators of the Encyclopedia of Life. To get it on its way and develop it fast, some argued that it requires input from a broad community. The Encyclopedia of Life should be modeled on Wikipedia, to which a worldwide community of users can contribute. Others argued that access to the Encyclopedia of Life should require a port of entry that exercises scientific scrutiny. To have available information on plant and animal species reliably and meaningfully stored, species should be arranged in a phylogenetic system that reflects their evolutionary relationships. "Nothing in biology makes sense except in the light of evolution," the renowned population geneticist, Theodosius Dobzhansky, once said.[5] The philosophers of biology Kim Sterelny and Paul Griffiths[6] paraphrased this famous line as "nothing makes sense in biology except in the context of its place in phylogeny," that is, in the context of its place on the Tree of Life.

As the debate heated up, a proponent of the alphabetical arrangement rejected the need for phylogenetic information content in the Encyclopedia of Life. Indeed, the US National Park Service manual calls upon some "recognized taxonomic authority" on which the evolutionary arrangement of collections would have to be based. But who is to provide such authority if it is true, as one of the participants in the discussion once remarked that "90% of our current science is wrong anyway, and will have to be revised in the future"? This, of course, is a very strong statement, revealing a very skeptical attitude toward science. It is certainly true that hypotheses about evolutionary relationships of species are subject to change as new information becomes available. Science is supposed to be a self-correcting system, its theories subject to rejection or revision in the light of new evidence. That is what separates a scientific theory from a dogma. But to call contemporary science wrong by 90% is surely much too strong. If contemporary evolutionary biology is

---

[4]Handling and care of dry bird and mammal specimens. Conserve O Gram, September 2006, Number 11/9.

[5]Dobzhansky, T. 1973. Nothing in biology makes sense except in the light of evolution. The American Biology Teacher, 35: 125–129.

[6]Sterelny, K., and P.E. Griffiths. 1999. Sex and Death. An Introduction to Philosophy of Biology. The University of Chicago Press, Chicago, p. 379.

wrong by 90%, why should the doctrine of Intelligent Design not be respected as a serious alternative to natural selection?

1.2   Natural Theology and the Doctrine of Intelligent Design

The main focus of the doctrine of Intelligent Design is what its proponents call "irreducible biological complexity." Harking back on Wiliam Paley's old "watch-maker analogy" as expounded in his notorious *"Natural Theology"* of 1802,[7] the proponents of Intelligent Design highlight biological structures such as the bacterial flagellum, which is thought to reveal a mechanistic complexity so sophisticated that its evolution cannot possibly be satisfactorily explained on Darwin's theory of natural selection. "Intelligent Designers" portray the theory of natural selection as an example of a universal law of nature that – according to the teaching of the great philosopher of science Sir Karl Popper – must be considered falsified if only a single contradictory case can be convincingly documented. Karl Popper opened the sixth German edition of his famous book on *"The Logic of Scientific Discovery"* (of which much more later) with a quote from the immortal German philosopher Immanuel Kant: "The *modus tollens* is a form of rational argumentation, which concludes from effects to causes, and thus delivers proof not only very stringently, but also very easily. Because, if only one single false conclusion can be drawn from a statement, then that statement is false." The ultimate falsifying example high-lighted by "intelligent designers" is the flagellum, a filament of complex internal structure with which bacteria of a "pool" of species collectively referred to as *Escherichia coli* propel themselves forwards. However, questions that are elegantly glossed over in this account, and that will be addressed later in this book are as follows: what is a universal law of nature, how can it be falsified, and does modern science – in particular modern evolutionary theory – in fact deal with such laws and their falsification?

Proponents of "Intelligent Design" do not name the engineer responsible for the flawless function of the flagellum in protozoans, regarded by some as simple organisms presumably nested within the root-system of the Tree of Life. However, it is easily appreciated that the "watchmaker analogy" barely conceals the one and only Divine Creator as The Cause of organismic complexity. And indeed, if scientific knowledge is doomed to remain woefully incomplete, why should an initial act of Divine Creation not be integrated into evolutionary theory to render that theory complete? Even Darwin, perhaps to temper the outrage of his unpre-pared readers, concluded his *"On the Origin of Species"* of 1859 with the remark: "There is grandeur in this view of life, with its several powers, having been

---

[7]Paley, W. 1802. Natural Theology: or Evidences of the Existence and Attributes of the Deity, Collected from the Appearances of Nature. J. Faulder, London

originally breathed into a few forms or into one"[8] – "breathed" by whom? On which basis do biologists reject an initial act of Divine Creation, which in itself would not necessarily invalidate the claim for the existence of a hierarchical order in nature that is accessible to empirical investigation? Even if Noah released animal species from his ark in the alphabetical order of their names, this would not invalidate the claim that the Plan of Creation, originating in the thought of a Divine Creator, contained a blueprint of a Great Tree of Life, where all plant and animal species slot into a hierarchically structured classification according to their "affinities." To decipher the "affinities" of organisms and to order them accordingly would be tantamount to reading God's mind in nature. This was, indeed, the opinion of one of the foremost experts in zoological systematics of his time, Louis Agassiz (1807–1873). Just 2 years before the publication of Darwin's "*Origin*," Agassiz published his famous "*Essay on Classification*"[9] in 1857, arguing that one of the noblest tasks of zoology is the discovery of the natural hierarchy into which all living beings can be sorted. But for Darwin, the "affinities" of species were causally rooted in common descent, whereas for Agassiz they were rooted in Divine thought. If a hierarchical classification can be discovered – rather than being invented – by biologists, then this hierarchy cannot be considered artificial, but must exist in nature: "To me [i.e., Agassiz] it appears indisputable, that this order and arrangement of our studies are based upon the natural, primitive relations of animal life." Yet this being so, "those systems, to which we have given the names of the great leaders of our science who first proposed them" could "in truth" represent nothing but "translations into human language of the thoughts of the Creator."[10]

How should we counter Agassiz' assertions? After all, hierarchical order might just be a brutal fact of nature that requires no further explanation, as Agassiz himself realized: "a system may be natural, that is, may agree in every respect with the facts in nature... but merely as the expression of a fact existing in nature – no matter how."[11] The historian of science, Mary P. Winsor, found it characteristic of Agassiz to hide this point in a footnote at the bottom of the page, as this insight might well have proven fatal for his theory of classification.[12] The reason is that Agassiz added an explanation to what he declared to be an empirical fact based on observation, while declaring in a footnote that the explanation he proffered was not necessary for the recognition of this fact. Likewise, Darwin found the hierarchical classification of organisms to be a fact of nature, amenable to empirical investigation, yet he again was not quite satisfied with leaving this fact alone. Instead, he

---

[8]Darwin, Ch. 1859. On the Origin of Species. John Murray, London, p. 490.

[9]Agassiz, L. 1857. Essay on Classification. Contributions to the Natural History of the United States, Vol. 1. Little, Brown & Co., Cambridge, MA.

[10]Agassiz, L. 1859. Essay on Classification. Longman, Brown, Green, Longmans, & Roberts, and Trübner & Co., London, p. 9.

[11]Agassiz, 1859, ibid., p. 8, Footnote 1.

[12]Winsor, M.P. 1991. Reading the Shape of Nature. Comparative Zoology at the Agassiz Museum. The University of Chicago Press, Chicago, p. 25; see also Rieppel, O. 1988. Louis Agassiz (1807–1873) and the reality of natural groups. Biology & Philosophy, 3: 29–47.

likewise added an explanation for its existence, which is descent with modification: "Descent being on my view the hidden bond of connexion which naturalists have been seeking under the term of the natural system."[13]

1.3   The Impact of Modern Philosophy of Science

Again it is Karl Popper who claimed to have shown the way of how to choose between the explanations of the natural system offered by Agassiz and Darwin, respectively. Popper's earliest claim to fame was his insight that there is no theory free observation. According to this thesis, which will require further discussion later on, brute facts of nature cannot be intelligible. But Darwin offered an explanation for the existence of the natural system that is grounded in natural causes, whereas Agassiz, explanation transcended the natural course of events by appealing to a supra-natural Creator. Popper thought that theories of science that concern natural processes or events should be testable, and potentially falsifiable through tests, whereas any explanations that transcend the natural course of events and for this reason are not testable must be considered metaphysical and hence be excluded from scientific discourse. Along such a demarcation line, Darwin's theory would qualify as scientific, whereas Agassiz' explanation would gain no purchase in scientific discourse because of its metaphysical character.

Popper's views on the philosophy of science became widely known and broadly accepted within the scientific community and the public at large. Scientific theories, he proclaimed, are those that can be tested by experiment, and as a consequence of such tests, it can potentially be found to be false. The converse, that scientific theories could be empirically tested, and potentially found to be true, was denied by Popper. Taking this position to its letter would mean that scientists can never know whether their theories are right, they can only know when and why they have gone wrong. This is the root of the claim that science is doomed forever to remain incomplete, and that 90% of contemporary science may be wrong and in need of revision in the future. Such a highly skeptical view of science became popular in the 1960s and 1970s, mainly as a reaction against the logical positivists' defense of the possibility to confirm scientific theories, and by some – including some highly respectable scientists – such skepticism is defended to the present day. As will be shown, it was promoted not only by Popper, but also – albeit in different ways – by three other very prominent philosophers of science, each with a large audience amongst the broader public: Thomas S. Kuhn, Paul Feyerabend, and – to some lesser degree – Imre Lakatos. Today, the history of the philosophy of science has turned another page or two. The battle against the logical positivists has been fought, supposedly won, and then found to have thrown out the baby with the bath water. A search is underway to clear former misunderstandings, and to give

---

[13]Darwin, 1859, ibid., p. 449.

empiricist philosophers proper credit where credit is due. As will be shown, Popper's authority has somewhat faded, and his strictly falsificationist attitude toward scientific theories has been recognized as internally inconsistent. The philosophical foundation on which Thomas Kuhn built his sociology of science has been revolutionized, and a new approach toward the justification of the undisputable success of science has been formulated. Science certainly cannot take possession of absolute truth, but it can – and does – do better in "tracking the truth"[14] than Popper, Kuhn, and Feyerabend would have admitted. A new generation of philosophers takes a different approach in the analysis of science and its success in space exploration, biotechnology, and medicine, to name just a few of its branches. Whether a universal law or not, whether falsifiable in a Popperian sense or not, natural selection theory can be – and is – used successfully in the development of computer software and vaccines against HIV.

1.4   An Outline and Some Historical Context

The story to be told, then, is how the science of biology came to free itself from claims of initial Creation or Intelligent Design. The story will also reveal that claims of initial Creation, or Intelligent Design, as they are issued today, are in fact an old hat. The relevant issues were hashed out first in France in the late eighteenth and early nineteenth century, later in Victorian England in the middle of the nineteenth century. In 1844, Robert Chambers anonymously published a book titled "*Vestiges of the Natural History of Creation.*" It presented nature not as a Tree of Life, but rather as a Ladder of Life, and argued that life had evolved along that ladder from mushroom to human, on the basis of purely natural causes, without any intervention from above. This book created an enormous controversy,[15] and was particularly opposed by Hugh Miller, who in 1849 published his "*Foot-Prints of the Creator or, the Asterolepis from Stromness.*" Miller hoped not only to expose the fatal flaws in Chambers' treatise, but also those of an earlier French version of an evolutionary theory, that of Jean-Baptiste Lamarck, which provided guidance to Chambers' writings. Miller's "*Foot-Prints*" was considered the ultimate refutation of transformationist ideas, until it was recognized that its interpretation of the Fossil Record was flawed.[16] Darwin naturally followed the controversy closely, learning from it which pitfalls to avoid in his own rendition of evolutionary theory, which he would publish in 1859, and which would be the one to catch on and change the world.

---

[14]Psillos, S. 1999. Scientific Realism. How Science Tracks Truth. Routledge, London.

[15]Secord, J.A. 2000. Victorian Sensation: The Extraordinary Publication, Reception, and Secret Authorship of Vestiges of the Natural History of Creation. The University of Chicago Press, Chicago.

[16]Secord, 2000, ibid., p. 282.

To enter into the debate between Chambers and Miller, we first need to find some historical context. It is necessary to understand the contemporary sciences of biology, geology, and paleontology (the study of fossils) to appreciate the impact of Chambers' book, and the force of Miller's rebuttal. With the Christian interpretation of Ancient Greek philosophers and naturalists, we enter a created world. We learn what the characteristics of such a created world are, a world that just "is," that never truly "becomes." It is a world governed by universal laws of nature, which themselves never change, just as the fundamental structure of the universe likewise never changes. It is true that, looked at from our vantage point, planets are in continuous motion, but they move on seemingly eternal, immutable orbits that can be described in the equally time-less language of mathematics. By contrast, a chicken embryo seems to undergo drastic changes in size, shape, and composition during its development. On the one hand, astronomers had no "problem of change." On the other hand, biologists developed ingenious theories to deal with change as is apparent in the developing embryo, desperately seeking to avoid the paradox that change creates in a created world.

However, time obtains from the passing of nature, as the philosopher Alfred North Whitehead so aptly observed, and with the passage of nature change crept into theories of biology after all. Offspring inherit a variable mixture of characteristics from their parents, and occasional malformations can happen: it was the study of development and regeneration of organisms that made it impossible for biology to deny change. Once this insight had become unavoidable, books such as those written by Lamarck and Chambers became possible. The dispute between Chambers and Miller turns much on geological, paleontological, and biological evidence. Major issues in the debate were the nature of the Fossil Record, and whether the succession of fossilized forms of life through geological time supported Chambers' vision. Evolution is a process of change, and evolutionary theory is an explanation of the causes that drive this process of change. But for an evolutionary process explanation to be applied, we first have to have a pattern of a natural order in need of an explanation. Systematists classify organisms: it was the eighteenth century Swedish botanist Carl von Linné (generally known as Carolus Linnaeus) who brought methodological rigor and strength to this discipline. But if we classify organisms, is the resulting classification purely logical in nature, as Miller thought, or does it reflect order in nature, as Chambers claimed? Is the classificatory system created in the classifiers mind, or is it discovered through the study of nature? And if order in nature is discovered rather than created by systematists, what kind of order is it? Does such a natural system correspond to the metaphor of a ladder, as Chambers thought, or to the metaphor of a branching tree, as Darwin thought? As embryos develop from the fertilized egg to the hatchling and to the sexually mature adult, they gain in complexity. Is it true that embryonic development follows along the ladder of life, as Chambers thought, and that the Fossil Record again mirrors this ladder of life, delivering a threefold parallelism of the natural system with embryonic development and the succession of fossils through layers of rock? The ladderized natural system, embryonic development, and the Fossil Record were all thought to display the same pattern of progress toward greater complexity of life forms, and it is this pattern that Chambers explained through his "Law of Development." Miller used the then oldest known fishes of curious yet highly complex structure that he had dug up in the Old Red Standstone of Scotland to highlight apparent imperfections of the threefold parallelism and on that basis to refute its causal explanation through the hypotheses about natural causes that Chambers had offered.

But Miller not only confronted Chambers on the basis of the evidence the latter had adduced in support of his vision but he also shifted the debate to a philosophical level. What is respectable science? How should science be properly pursued? How is science to be organized, and what are the appropriate methods for the various scientific disciplines? What are "laws" of nature, and how should a proper scientific explanation be structured? What can, and what cannot, count as relevant evidence? What role does observation play in science, and how far can a scientist legitimately push his/her speculations before losing credibility?

If all this sounds familiar, it is because the very same questions are on the table again today, with respect to Creation Science and the doctrine of Intelligent Design. This is the point at which Karl Popper, Thomas S. Kuhn, Imre Lakatos, and Paul Feyerabend come in. Popper believed in the unity of the scientific method, but through the method he championed he rendered science incapable of obtaining any positive knowledge. Kuhn thought that science is a product of its historical and socio-political context. In different historical and social contexts, different scientific research programs will be guided by different theories. According to Kuhn – or at least according to some interpretations of his writings – the scientists of a given period construct the world they claim to discover, at least in part, through their own theories. When scientific theories change, the world of scientists changes with them. Imre Lakatos sought a synthesis between Popper's falsificationism and Kuhn's sociology of science through the introduction of his concept of institutionalized research programs. Finding much he liked in Kuhn's philosophy, Feyerabend finally took the socio-political approach to science one crucial step further forwards, declaring that there is really no such thing as a proper scientific method. Well, if that is true, how are we to judge evolutionary theory against Creation Science?

We will see what reasons there are to conclude that Popper, Kuhn, and Feyerabend drew a far too pessimistic picture of science, a picture that can certainly be much improved upon. But it will also become clear that when they talked about science, Popper, Kuhn, and Feyerabend took physics as the paradigmatic example of a natural science, and that was it. Ironically, it is also from physicists and astronomers that Darwin earned the most serious objections to his theory of evolution at the level of a scientific (as opposed to a social, moral, or theological) debate. Darwin characterized his "principle of natural selection" as "one general law, leading to the advancement of all organic beings, namely multiply, vary, let the strongest live and the weakest die."[17] However, the astronomer John Herschel objected that his "law" did not qualify as one of those universal laws of nature that a proper branch of science

---

[17]Darwin, 1859, ibid., p. 244.

seeks to discover. Darwin spoke of the "laps of ages" it took for species to evolve, but the physicist Lord Kelvin delivered putative "mathematical proof" (now known to have been erroneous) to show that the earth was too young to accommodate Darwin's theory of evolution. However, biology, and especially historical biology, is not physics. With Darwin, evolutionary biology became an autonomous scientific discipline, with different methods and different schemes of argumentation than those in physics. The reception of Darwin's theory in continental Europe makes this particularly clear: "While it is true that biology has to continue its development as a science about natural laws, it is also true that biology cannot be only a science about natural laws. This is because research that involves organisms concerns not only lawful, but also historical relations. In that sense, biology differs from physics and chemistry both in method, as also in the scope of inquiry."[18] Even if the arguments of Popper, Kuhn, and, perhaps to a somewhat lesser degree, those of Feyerabend gain some purchase with respect to physics, they do not do so in the same way with respect to historical biology. Popper claimed science to progress through conjecture and refutation. Theories are conjectured, then tested against experience, and rejected if they fail the test. Darwin, in contrast, set out to trace the traces that evolutionary history had left behind in the Fossil Record as much as in the living world.

---

[18]Uhlmann, E. 1923. Entwicklungsgedanke und Artbegriff in ihrer geschichtlichen Enstehung und sachlichen Beziehung. G. Fischer, Jena, p. 111.

Chapter 2
The Problem of Change

An evolving world is a world of change. A created world does not change. It just is. Or if it seems to change, the change is only apparent, as it is preconceived and preordained by the blueprint of Creation. Change is paradoxical: how can something change and yet remain the same? How much remodeling can be done to a house before we no longer call it the same house, but a new and different one? Some Ancient Greek philosophers solved the 'problem of change' through the concept of dynamic permanence: planets are in constant motion, continuously changing their position relative to other heavenly bodies, but they travel in immutable, eternal orbits. These orbits can be described in terms of universal laws of nature, which in turn can be expressed in the timeless language of mathematics.

The concept of dynamic permanence is less easily applied to organisms. The developing chicken appears to change continuously, but here, organs such as the heart, the brain, and the limbs seem to come into existence without having been apparent before. Pre-evolutionary biologists solved the 'problem of change' through the doctrine of pre-existence. The entire organism pre-exists since the time of Creation, folded up into minute dimensions, and encapsulated either in the spermatozoon's head portion, or in the female egg. Embryonic development, a process of change, then becomes the mere unfolding of structures that are preformed and that already pre-exist since the beginning of time, albeit too small to be seen. Such an unfolding of pre-existent structures during the development and growth of the embryo was the original meaning of the term 'evolution'.

2.1 Change in a Created World

"Evolution" literally means "unrolling" or "unfolding." Something that is unfolding is something that takes part in a process, or that is a process itself. A process, in turn, is a chain of events that naturally stretches through time. Evolutionary theory is about natural processes that extend through time and result in change. An evolving world is an ever-changing world. This contrasts with a created world that does not change. A created world is just the way it was created. Nothing comes, nothing goes, and everything remains the same always. Any apparent change is just what it is – apparent, that is preconceived and preordained through the blueprint of Creation. Can something come from nothing? Can something dissolve into nothing? What is it

O. Rieppel, *Evolutionary Theory and the Creation Controversy*,
DOI 10.1007/978-3-642-14896-5_2, © Springer-Verlag Berlin Heidelberg 2011

to say that "something changed"? It is to say that one and the same thing underwent change? But how can anything change and yet remain the very same thing, rather than become something else as a consequence of change? This is the paradox of change. Creation is not changing one thing into another. Instead, it puts things into place. Things are put into space and time through Creation. Thereafter, they stay the same and remain unchanged.

If the world is an evolving world that undergoes constant change, the Creator cannot be of this world. Change implies time, and time obtains from the passing of nature, as the philosopher Alfred North Whitehead puts it: "There is no holding nature still and looking at it."[1] But the Creator is eternal and timeless. Neither does He come out of the future, nor does He recede into the past. He resides outside time and space, so to speak. To enter into an evolving world would mean to enter into change, to enter into time, to abandon timelessness, and to abandon eternity. And yet, at some point, at the beginning of all things, the Creator seems to have undergone change, when He enacted His Creation. But to say that the Creator is eternal, existing beyond time, and to also say that the Creator entered into time with His Creation, is a logical contradiction. The Creator cannot reside outside time and enter into time at one and the same time, that is, at the beginning of all things. Such a logical contradiction is nonsensical. To speak in logical contradictions makes no sense, it means to say nothing.

A created world therefore cannot change, and if change appears to occur to the human observer, this cannot be real change. Human powers of perception are limited to processes that extend through time and space. The apparent change that takes place in the world must be a mere impression of change, an illusion the human observer takes away from his/her fleeting, every-day observations. The underlying structure of the universe must remain the same. The universe is timeless, eternal, governed as it is by equally eternal, timeless, and universal laws of nature that constitute the blueprint of Creation. This is a very comfortable world to live in. It is a secure world, a world one can know something about. A world that functions like a universal clockwork can be explained in terms of its underlying machinery that never changes. This world is one the past of which can be known, and in which the future can be planned accordingly. If the fundamental structure of the universe is eternal, if the laws governing the universe are eternal too, then knowledge of these laws translates into knowledge of the universe. Observed effects can be explained as a consequence of their cause, the past becomes explicable, the future becomes predictable. It is only the very beginning that remains unexplained. That's where the paradoxical Creator comes in, and with Him the Plan and Purpose of Creation.

So what does this Plan of Creation look like? Well, there seem to be fire and air, water and soil. There seem to be rocks and minerals, plants and animals. Some living organisms look almost like rocks, as do some lichens and mushrooms. Some animals look almost like plants, as do some sea anemones and sponges. Worms

---

[1]Whitehead, A.N. 1920. The concept of nature. Tarner lectures delivered in Trinity College November 1919. Cambridge, Cambridge University Press, UK, p. 14.

appear to be lowly creatures in comparison with fish. Fishes appear to be lowly creatures in comparison with parrots and tree shrews. Lions appear more sophisticated in their pursuit of prey than salamanders. Sea gulls appear more sophisticated in their care for offspring than snakes. At the top of this ladder of life sits the human being, distinguished from all other living beings by the capacity for rational reasoning, and by the capacity to communicate such reasoning through language. The Creator resides outside Nature; angels provide the link between Him and humans. A thread seems to run through this Creation, linking lower with higher forms of life.

The direction from bottom to top in the ladder of life is not meant to imply change, however, even if its rungs appear to have been put into place at different times during the history of the planet earth.It is, instead, a static arrangement of life in a Great Chain of Being that reflects the immutable Plan of Creation.[2] The concern in laying out the Plan of Creation is not change, but the order that pervades nature instead. The Great Chain of Being was not related to a historical process of the emergence of increasingly complex forms of life, where something genuinely new successively evolves from the old. It was meant to express a hierarchical classification of the contents of nature, according to the Book of Creation. Eighteenth century French aristocrats extended the ladder of life from the realm of nature into the realm of society. They justified the social hierarchy of human society as a corresponding expression of the eternal Plan of Creation: firm, immutable, safe, and secure. French biologists of the eighteenth century, who were part of this aristocracy, or at least maintained close ties to its members, chastised speculations about genuine change in nature as infected with the seed of atheism. They went to great length to explain away any apparent changes in nature to provide the doctrine of an immutable, eternal, Great Chain of Being with a scientific foundation.[3] Theories departing from that doctrine were castigated as godless, immoral, and subversive, and their proponents prosecuted. But change prevailed: The French Revolution swept away the ancient social order, and the theory of evolution swept away the doctrine of a static Great Chain of Being.

Later still, time was found to be relative, matter was found to be a form of energy, and quantum mechanics showed the basic structure of the universe to be indeterminate, and the laws of nature to be probabilistic. The past still is explicable, the future still is predictable, but not with absolute accuracy nor with a certainty that is rooted in the eternity of Creation. The blueprint for Creation, if ever there had been one, would have been blurred. We are stuck in a logical contradiction again: the Creator cannot be almighty and omniscient, yet undecided and ambiguous. Albert Einstein, who wanted to unlock the ultimate secrets of the universal clockwork, rejected such fundamental indetermination of the world, as was proclaimed by quantum mechanics. God, he insisted, "does not play dice" with the universe.

---

[2]Lovejoy, A.O. 1936. The Great Chain of Being. Harvard University Press, Cambridge, MA.

[3]For a good illustration of these tensions see Sonntag, O. 1983. The Correspondence between Albrecht von Haller and Charles Bonnet. Huber, Berne.

The quantum mechanics community replied: "Einstein, stop telling God what to do with his dice."[4]

How could all these changes happen? Why and how did science try to avoid the problem of change by explaining change away? And how did it come about that change became a dominant perspective of the modern worldview? To trace these changes, and with them the tension between the Creator and the passing of nature, is the topic of this book.

2.2   The Problem of Change

Western traditions in philosophy and natural science originate with Ancient Greece. Philosophy investigates the world of thought, and natural sciences investigate the world of matter. Natural sciences strive to acquire explanatory knowledge of the material world and of its inhabitants. One of the wonders of this world is the harmonious movement of the stars in the nightly sky. Fascinated by the stars since ancient times, people have been observing the nightly sky, studying the regular movements of celestial bodies. This is a most fascinating enterprise indeed. The nightly sky seems forever to be subject to continuous change: the sun rises in the morning and settles in the evening; the moon comes and goes with a regularity that parallels oceanic tides. Time passes as changes occur and reoccur. And yet, the language of mathematics, which itself is timeless, eternal, and universal allows us to describe those regular and recurrent changes in the nightly sky with great precision, revealing them to be governed by the universal laws of nature. But how can change be described in a language that neither knows past, nor future tenses? The bright star that outshines all others after sunset in the evening sky is the "Evening Star," called Hesperus by the Ancient Greeks. His "brother," the star that appears before the sun rises in the morning sky is the "Morning Star," called Phosphorus by the Ancient Greeks. It was the great philosopher and scientist Pythagoras who recognized that Hesperus and Phosphorus are, in fact, not two different stars, but one and the same "star" instead, that is, the planet Venus. Some authors say that this was discovered earlier, by Babylonian astronomers. Either way, while the human observer perceives continuous changes in the sky, the coming and going of seemingly different stars, there is, in fact, no change. The observed phenomena reveal nothing but the "becoming visible" of one and the same, numerically identical planet at different times, in the East in the morning, and in the West in the evening. The planet Venus is the same at the present time as it was in the past, and it will remain the same in the future. It does not change. It moves in an orbit that likewise does not change and that, for this reason, can be described in the timeless language of mathematics. It is only from the vantage point of the human observer that change seems to occur: the

---

[4]http://www.bbc.co.uk/sn/tvradio/programmes/horizon/einstein_symphony_prog_summary.shtml (accessed 12/28/07).

"morning" star seems to come and go, as also does the "evening star." But this is merely apparent change in an unchanging world. Venus does not change, nor does the orbit along which it moves. This world, which some interpreters trace back to Aristotle amongst Ancient Greek philosophers[5], is one of the dynamic permanence: the observable movements, the changes, the dynamics that seem to permeate the observable world are ultimately reducible to the unchanging, that is, static and fundamental structure of the universe, governed by timeless, universal laws that can be expressed in the equally timeless, universal language of mathematics. To sketch a world of dynamic permanence represents an attempt to identify a unified and unchanging "being" that underlies the continuous "becoming" of an ever changing multiplicity of phenomena.[6]

The ancient Greek philosopher Aristotle is famously known as the "father of biology," but also as one of the founders of formal logic. Could it be possible, or even meaningful, to describe the development of a chicken embryo in the language of formal logic? Aristotle, like many philosophers before – and after – him, struggled with the "Problem of Change," and the paradox it creates. How can something change, and while changing, how does it still remain one and the same "something"? Change implies the existence of an object that undergoes changes in its properties. And yet, all the while it changes, this object must remain the same, numerically identical object if it is to be the case that it is *this* object that undergoes change. The prominent twentieth century philosopher of science, Sir Karl R. Popper concluded: "every change is the transition of a thing into something with, in a way, opposite qualities."[7] The problem of change leads us into a paradox again. Consider an acorn that develops and grows into an oak tree: the descriptions of the acorn and of the oak tree are vastly different, and yet they apparently apply to one and the same individual organism. Or look at the caterpillar that transforms itself into a pupa from which a butterfly eventually emerges: is it possible to say that the caterpillar and the butterfly is one and the same, self-identical thing? And if it is, how are we going to express this in the language of logic? Ancient Greek philosophers are known for their love of paradoxes. Here is one, the famous ship owned by Thales of Miletus: constantly exposed to wind, weather, and waves, one plank after another is removed from it as necessary, and replaced with a new one. Intuitively, it seems right to call it the same ship, that is, the ship of Thales, even if after several years all of its original planks had been gradually replaced by new ones. But now consider a mischievous philosopher who takes away one plank after another, and slowly, again over several years' time, replaces the removed planks with new ones. But as he continues his repair work, he secretly stores the removed planks in a hidden place, and at the time when all the planks of the ship he started out with have

[5]Balss, H. 1943. Aristoteles biologische Schriften. Ernst Heimeran, Munich.

[6]Uhlmann, E. 1923. Entwicklungsgedanke und Artbegriff in ihrer geschichtlichen Entstehung und sachlichen Beziehung. G. Fischer, Jena, p. 14.

[7]Popper, K.R. 1989. Conjectures and Refutations. Routledge & Kegan Paul, London, p. 144f.

been replaced, he rebuilds from the planks he had squirreled away the original ship. We now have two identical ships – which one is Thales' ship?

One way to deal with paradoxes is through logical analysis, so let us look at the "problem of change" from a logical point of view. Given the appropriate axioms, mathematical theorems are considered to be universally true. Simply put, it is inconceivable that there would have been a time in the past, or that there would be a time in the future, at which "3 + 3 = 6" is false, whereas "6 – 3 = 5" would be true. Mathematical journals that are published in China and America use the same formal language to express the mathematical relations. The same is true for the language of logic, which again is timeless and universal. Just as there are timeless truths in mathematics, so there are timeless truths in logic, which can be formulated in terms of universal laws of logic.

One such basic law of logic is the law of noncontradiction: it says of two contradictory propositions that it is impossible for both of them to be true. Only one of those propositions can possibly be true; one must necessarily be false. But now consider the following proposition:

"This apple is all-over green *and* this very self-same apple is all-over red."

If the apple is all over-red, then it is not all-over green. So we can reformulate:

"This apple is all-over green *and* this very self-same apple is not all-over green."

But this, according to the law of noncontradiction, is false. One and the same, self-identical apple cannot be all-over green *and* all over-red. Of course, we could say that yes, one and the same apple cannot be all-over green and all-over read *at the same time*, but that one and the same apple can turn from being all-over green to being all-over read *through time*, that is, as it ripens. But the law of noncontradiction cannot accommodate change through time; there is no temporal dimension to that law. It is true that "being all-over green" is a property of an apple, as is "being all-over red", such that it might seem possible to anchor these properties in time: one and the same apple is all-over green in early summer, all-over red in the fall. But that argument avoids, rather than solves the problem of change, for the apple no longer undergoes genuine change. It just comes to relate to different properties at different times, a view of things bound to create metaphysical nightmares.[8] To overcome the problem of change, the Greek philosopher Parmenides concluded that there simply cannot be any real, genuine change.[9]

2.3 The Distinction of Essential and Accidental Properties

Aristotle saw things a little bit differently. When we say that it is one and the same apple that changes from being green to being red over time, then to "be green", or to "be red", cannot be properties in which to anchor the self-identity of that changing

---

[8]Heil, J. 2005. From an Ontological Point of View. Oxford University Press, Oxford.

[9]Mortensen Ch. Change. The Stanford Encyclopedia of Philosophy (Winter 2006 Edition), Edward N. Zalta (ed.), URL = http://plato.stanford.edu/archives/win2006/entries/change/.

apple. There must be something else, something other than its color, which establishes the self-identity of the apple through time and change. Aristotle tackled the problem of change by introducing a number of basic distinctions. First, he distinguished between material objects that extend through space, and properties that are not substantial but that are exemplified by objects. The apple is a substantial object that exists in time and space: we can pick it up and look at it from all sides before throwing it away or eating it. The properties of "being green" or "being red" are not substantial, but instead are exemplified by the apple at different times. We cannot pick up "being green", nor can we kick away "being red." Among properties, Aristotle further distinguished essential, permanent, unchanging properties from accidental ones, that is, properties that an object can gain or lose over time. An essential property is one that any object cannot fail to exemplify under any circumstances in which the object exists. An accidental property is one that an object can shed or acquire. Venus appears to be subject to constant change as it appears and disappears in the morning sky and in the evening sky. Such changing properties, that is, "appearing in the morning sky," and then again "appearing in the evening sky," must be accidental properties of Venus. But these observable changes that Venus undergoes reoccur with remarkable regularity, one that can be precisely calculated. The reason is that Venus orbits the Sun (to the Ancient Greek, it seemed to orbit the earth) along a unique yet immutable trajectory. To travel on this never changing trajectory is an essential property of Venus. For Aristotle, the self-identity of a substantial object through time and change was grounded in its essential properties, that is, in properties that are permanent and that cannot change.

If the apple changes from being green to being red over time, then these colors cannot be essential properties of the apple. These must be accidental properties of the apple, instead. There must, therefore, be some other, essential, property that belongs to the apple and thus allows us to claim that it is the very self-same apple that changed from green to red. This property is the numerical identity of the apple that undergoes change: it is the numerically identical object that is once green, later red. If a caterpillar turns into a butterfly, and if both are said to be one and the same organism, then both the caterpillar and the butterfly must share some property that did not change during metamorphosis, and this property would be the essential property of that organism, establishing its numerical identity through change. The colors of the caterpillar or of the wings of the butterfly are accidental properties of those organisms, ones that apparently can undergo change. So what would be the unchanging essential property shared by the caterpillar and the butterfly that develops from it? It is the unique origin of that individual organism from one-particular fertilized egg that establishes its numerical identity. The same holds for the apple that changes from green to red: it is its origin from one unique fertilized flower that establishes its numerical identity. In an evolving world, a species can undergo change while remaining the same species. Its numerical identity through time and space is anchored in its unique evolutionary origin. In an evolving world, an ancestral species lineage can split into two descendant species lineages. Making two out of one means that an ancestral species gave rise to two numerically different descendant species. The same cannot be possible in a created world, for in such a world, there is no unique evolutionary origin for a species, as there also is no origin of two new species from an ancestral one. In a created world, it is the unique act of creation of a species that anchors its identity through generations. Offspring differ from their parents in many ways, but all these differences could only be accidental differences. In a created world, the species, to which both parents and offspring belong, cannot undergo any essential change. The extinct mammoths look different from the living elephants. One could imagine that mammoths changed over time to give rise to the living elephants. But in a created world, the change from a mammoth to an elephant could only be apparent change, a change in accidental properties that was preconceived in the blueprint of Creation. If the elephant lineage is genealogically connected to the mammoth lineage, they must both belong to the same species, created at the beginning of time. They would have to share some underlying essential property, not visible to the human observer, which preserved the self-identity of the species through time and change, and that property would be the initial creation of the mammoth–elephant lineage.

A process that brings about change is composed of a series of events that stretches through time. For Aristotle, an event of change occurs if one and the same substance exemplifies one accidental property at one time or during one period of time; another accidental property at another time or during another period of time. Venus travels along an orbit that remains essentially unchanged, but becomes visible at different times at different horizons. That way, Aristotle obtained a world of dynamic permanence: everything is in motion, everything is changing all the time, but behind that apparent change lies the permanence, the timeless uniformity of nature, governed by timeless and uniform Laws of Nature that fix the orbit of Venus.

2.4   Embryos and the Problem of Change in Organisms

It is, perhaps, easy to comprehend that planets circle the sun on eternal orbits that are determined by timeless and universal laws of nature. However, it is less easy to comprehend how organisms that are born, develop, grow, and eventually die and decay would likewise circle through a world of dynamic permanence that is governed by universal laws of nature. But this is none-the-less the way early biologists understood the nature of species before evolutionary theory changed biology: the acorn grows into an oak tree, and the fertilized chicken egg develops into a chicken. These are processes of apparent change. But the oak tree species (e.g., the species *Quercus alba*), or the chicken species (e.g., the species *Gallus gallus*), do not change. They remain the same, forever. With a book published in 1651, William Harvey[10] – otherwise known as the discoverer of blood

---

[10]Harvey, W. 1651 [1981]. Disputations Touching on the Generation of Animals. Translated with Introduction and Notes by G. Whitteridge. Blackwell, London.

circulation – followed up on Aristotle's studies by opening chicken eggs to watch the embryo develop. Initially, there seemed to be a sort of a white, milky streak on the surface of the yolk, in the midst of which would gradually appear a little red, pulsating dot that became known as the "salient point" (*punctum saliens*) – the heart. From it would grow red filaments – the first rudiments of the blood-vascular system. Folding of tissues would later become observable – the chicken embryo starting to look like a maggot. Still later, head, trunk, and tail would become discernable. Tiny limb buds would eventually grow out of the trunk, and so on. The same process unfolded in a seemingly identical way in all the chicken eggs that Harvey examined, a regularity that suggests the developmental process to be governed by universal laws of nature. During its development, the embryo shows a number of changes of properties in terms of its size, shape, and composition – but could there be real change involved? Or are these just changes in accidental properties, and if so, what are the essential properties of animals and the species to which they belong? Harvey thought that the immutable, essential property of the developing chicken is its soul. It is this immortal soul that guides the development of the chicken to its species-specific goal according to a purpose, and this purpose is to perpetuate the species and its perfect adaptation to the place in the household of nature to which it has been assigned by its Creator. Modern biologists might be tempted to say that the essential, immutable property of the developing chicken is its genotype, but the analogy is misplaced. Genetics teaches us that the genome changes in the course of sexual reproduction, as the genes of both parents get mixed up and recombined in the genotype of the offspring. This is, after all, the stuff of evolution, a process that knows no goal or purpose. The soul, in contrast, is an essential property of the developing chicken that never changes. It is eternal and immortal. For Harvey, the soul was the guiding principle of embryonic develop-ment: it guided the development of the chicken in a purposeful way toward a well-defined goal: the persistence of its species through replication. The embryo seems to change during its development, but the species to which it belongs never changes. It is like the Morning Star and the Evening Star coming and going, while Venus remains immutably fixed on its orbit. Chickens develop, hatch, mature, reproduce, age, and die – they come and go. But the species remains the same, perfectly fulfilling the role assigned to it in the Plan of Creation. So what looks like change on the surface is not a real, essential change after all; it is only a change in temporary appearance, change in accidental properties. The fundamental structure of the universe remains the same, always, and forever.

But should the heart and the blood, the brain and the nerves that successively appear during its development really be considered as mere accidental properties of the developing chicken? Accidental properties can change: the heart of a fish has two chambers, while that of a chicken has four chambers. The brain of a fish looks very different from the brain of a chicken. But according to Harvey, the soul of a fish larva makes sure that it develops according to the blueprint that defines its species, just as the soul of the chicken guarantees the immutability of the blueprint that defines its species. A four-chambered heart looks more sophisticated than, and hence superior to a two-chambered heart, but there is no implication that one changed into the other. They remain the same, forever, representing different rungs in the ladder of life, different steps in the Great Chain of Being. The early chicken embryo seemed at some stage of its development to resemble a maggot, the latter representing a much lower rung in the ladder of life than a bird. At its first appearance, the heart of the chicken is a simple pulsating vesicle, simpler yet than the two-chambered heart of a grown fish. So during its development, according to Harvey, the chicken embryo seemed to climb up the steps of the Great Chain of Being. But that, to him, was no evidence of essential change, and no evidence of progressive development. Instead, the development of the chicken merely expressed the same great Plan of Creation over again: the same hierarchical order that characterizes the Great Chain of Being also characterizes the embryonic development of the chicken. Harvey found embryonic development to run parallel to the Great Chain of Being, but the parallelism is one of a static order, not one of dynamic change. Later authors recognized that fishes with their two-chambered heart were put into space and time at a different epoch of earth history than birds with their four-chambered heart. But that for them again would not imply species change. The species remain the same, designed according to the blueprint of Creation. They just become apparent at different times and in different places in a succession that mirrors the Great Chain of Being once again.

And yet, it is easy to say that an apple undergoes accidental change as it changes from being green to being red. Such a change of color in an apple does not map on the changes that occur during the development of a chicken. The apple merely changes in color. In contrast, parts of the organisms seem to come into existence during the development of a chicken that had not been there before. There was no sign of a heart before the appearance of the little red, pulsating dot. There was no sign of a brain before the folding of tissues that resulted in the formation of a head, trunk, and tail. Genuine change seemed indeed to occur in the development of a chicken. There is not merely a persisting substantial object that changes its colors or other accidental properties. Instead, there are new substantial parts of an organism, a heart, a brain, which seem to come into being. It is for such reasons that Harvey's 1651 account of embryonic development left some fellow biologists unconvinced or at least unsatisfied. The heart made its appearance early during development, but where did it come from? The vesicles that would form the brain appeared after the pulsating little red dot, but where was the brain before it started to become apparent? Many biologists and philosophers of the seventeenth and eighteenth century looked back on Aurelius Augustinus, known today as St. Augustine, for a solution,[11] for he, in the fifth century, famously dealt with another paradox of change, which was the resurrection of Christ into the realm of Deity. Being born from a human, and suffering death, Christ assumed a divine nature through His resurrection. How could this be possible, given that His divinity is an essential property of Christ, a property that transcends space and time, one that is eternal and

---

[11]Roger, J. 1971 Les Sciences de la Vie dans la Pensée Française du XVIIIe Siècle, 2nd ed. Armand Collin, Paris.

immutable? An eternal property cannot come into being at a certain location in space and at a certain point in time, for if it did, then it would not be eternal. To solve the paradox, Aurelius Augustinus invoked his famous doctrine of preexistence. Just as the oak tree preexists in the acorn, he said, so did divinity preexist in Christ, but it became actualized only upon His death and resurrection. Taking their clues from Augustine, biologists set out to develop perplexingly complex models of animal reproduction to avoid the paradoxical problem of change. The essential property of an animal species was rooted in its initial Creation at the beginning of time: its essence maintains the perfect adaptation of the species to the particular place in the household of nature to which it had been assigned by the Creator. If embryonic development of an organism seemed to string together a sequence of profound changes, this could only amount to a mistaken impression that resulted from the imperfect powers of human perception. Instead, development was explained as nothing more than a mere process of unfolding, a process through which preexistent structures became actualized, that is, functional. Augustine had claimed that the oak tree preexists, in its entirety but folded up and thereby reduced to miniature size, within the acorn.[12] As it develops and grows, it does not undergo any genuine change. Its growth, its development, is merely an unfolding and becoming apparent of what already exists. The wings of a butterfly preexist inside the caterpillar. They do not develop from rudiments, but merely unfold, as the butterfly emerges from the pupa. The world of biology is thus fully brought back into dynamic permanence: no real change would ever take place; what might appear as change to the human observer was merely the unfolding, the "becoming visible" of preexisting structures that had existed, encapsulated within one another, since the initial act of Creation at the beginning of time.

Antony van Leeuwenhoek was a Dutchman, who built the first microscopes, thus obtaining the means to observe things never seen by anyone before. His lenses drew a strange, vast, and varied microcosm close enough to be studied in detail. He communicated his discoveries to the Royal Society of London in a series of letters, which were then published in the prestigious "Philosophical Transactions" – the world's oldest scientific journal, which is still in existence.[13] In a letter drafted in 1677[14], for the first time in the history of biology, he described male spermatozoans that he had obtained from animals such as sturgeons and dogs. He believed that these little wiggly organisms carry the encapsulated, preexistent embryo into the female egg upon fertilization. The female egg would provide nothing more but the nutritive environment required for the development of the embryo, which would be the mere unfolding of the preexistent structures. Since the spermatozoon was able to propel itself forward with its tail, it had to be an animated animalcule, which is

---

[12]Augustinus, A. 1961. De Genesi ad Litteram Libri Doudecim. Translated by J.C. Perl. Ferdinand Schöningh, Paderborn.

[13]http://rstl.royalsocietypublishing.org/

[14]Leeuwenhoek, A. 1677. Observationes D. Anthonii Lewenhoeck, De Natis E Semine Genitali Animalculis. Philosophical Transactions of the Royal Society of London, 12: 1040–1046.

endowed with an animal soul. Since Leeuwenhoek believed the preexistent embryo to be encapsulated within the head portion of the spermatozoon, the latter also carried the soul of the embryo. He reported that huge numbers of spermatozoans are found in the seminal fluid of male sturgeons, dogs, and humans, and in the fact that only a tiny minority of those would likely go on to successfully fertilize an egg, many critics located a theological problem.[15] According to Leeuwenhoek's calculations, a huge number of souls would go to waste, something that was irreconcilable with the thinking of the time. In his *"Essay de Dioptrique"* of 1694, Nicolaas Hartsoeker claimed he had first observed spermatozoans "more than twenty years ago," thereby claiming priority in their discovery[16], a claim that turned out to be unsubstantiated, upon historical analysis.[17] However, Hartsoeker did publish the famous figure that shows a miniature homunculus fully preformed but folded up to a very small size in the head portion of a spermatozoan.

The same doctrine, called animalculism, was adopted by the great German philosopher, Gottfried Wilhelm Leibniz, who in his *Theodicy* of 1710[18] used it to establish the permanence of the relation between the body and the soul through death and resurrection. The development of the organism was a simple process of growth, a mere unfolding of structures that already preexisted; nothing new comes into being; development merely renders visible what already existed from the beginning of time. According to Leibniz, when the organism dies, it is folding up again to the dimensions of a physical point, too small for us to see. Resurrection is simply the reinitiation of the same process of unfolding. The attractive aspect of this theory is that the soul never needs to be stipulated to exist in separation from the body to which it belongs, something Leibniz believed to be impossible. Accordingly, the essence of each individual is its soul, and the changes an individual undergoes during development and ageing are changes in accidental attributes. But even these do not really change: what looks like change is merely the unfolding of preformed and preexisting structures.

However, some authors spotted troubles with Leibniz' theory. Leibniz had proclaimed that if an ordinary man falls asleep and suddenly wakes up again to find himself transformed to be the Emperor of China, he would have to remember his former life to appreciate the benevolence of God, who overnight transformed him to take up such a privileged position. Similarly, one would have to remember one's life to be able to appreciate God's benevolence at one's Last Judgment. But now consider the pious soldier who, through an unfortunate accident, is decapitated in battle. He dies and contracts to an invisibly small physical point. As he unfolds upon his resurrection, his body will be missing its head, and with it the brain (the seat of the soul), and with it the memory of his

---

[15]See discussion in Roger, 1971, ibid., esp. pp. 317ff. See also Hankins, T.L. Science and the Enlightenment. Cambridge University Press, Cambridge, UK, p. 135.

[16]Hartsoeker, N. 1694. Essay de Dioptrique. Jean Anisson, Imprimerie Royale, Paris, p. 222

[17]Roger, 1971, ibid., pp. 299ff.

[18]Leibniz, G.W. 1710. Essais de Théodicée. Isaak Troyrl, Amsterdam.

former life. The poor man would be called to his Last Judgment without the capability to comprehend the verdict.

2.5 Charles Bonnet and His Understanding of Evolution

The eighteenth century biologist and philosopher Charles Bonnet[19] from Geneva resisted his father's wishes for him to become a Lawyer, and instead took an early interest in insect life. Pursuing these interests, he discovered that captive female aphids, also known as plant lice, could produce viable and fertile offspring without ever having had any contact with the opposite sex. He separated aphids from the moment of their birth, keeping them in different powder jars. After just a few days, he found to his surprise that the isolated lice had multiplied. Charles Bonnet thus made history of biology with his discovery of parthenogenesis, the fertile reproduction by females of bisexual organisms without insemination by males.[20] For Bonnet, this discovery confirmed his friend Albrecht von Haller's conclusion that embryos were not preexistent in the male spermatozoans, but in the female egg instead.[21] It is the egg that contains the embryo, preformed, and preexistent in all its parts and ready to unfold, once the spermatozoon fertilizing the egg triggered that developmental process. In his native language, which was French, Bonnet called the process of the unfolding of preexistent embryos "*une évolution.*" This is the original meaning of the term "evolution."[22] The female of the first pair of each species created at the beginning of time would contain within itself eggs that encapsulated the germs of the next generation. Amongst the latter, the female germs contained the eggs that encapsulated the preexistent organisms of the next generation and so on – as far as the Creator had planned the natural course of events to unfold.[23] No change ever takes place, nothing new ever develops: nature works as small as is necessary to accommodate Divine thought. On Bonnet's account, the history of a species through time compares with the unpacking of a set of Russian nesting dolls. By the time of Bonnet's writing, it had been well-established that the surface of the earth had undergone dramatic changes through geological time. It had also become apparent that organisms preserved as fossils in successive layers of sedimentary rocks document the successive appearance and later disappearance of different species as one climbed up the quarry wall, breaking the rock with a

[19]Rieppel, O. 2001. Charles Bonnet (1720–1793), pp. 51–78. In: Jahn, J., and M. Schmitt (Eds.), Darwin & Co., Vol. 1. C.H. Beck, München.

[20]Bonnet, Ch. 1745. Traîté d'Insectologie. Première Partie. Durand Librairie, Paris.

[21]Haller, A. 1758. Sur la Formation du Coeur dans le Poulet; sur l'Oeil; sur la Structure du Jaune & c. Premier Mémoire. Marc-Michel Bousquet, Lausanne.

[22]Bowler, P.J. 1975. The changing meaning of evolution. Journal of the History of Ideas, 36: 95–114.

[23]Bonnet, Ch. 1768. Considérations sur les Corps Organisés, 2nd Ed. Marc-Michel Rey, Amsterdam.

hammer. All of this reduced in Bonnet's eyes to the unfolding of preexisting structures encapsulated within one another, according to the Plan of Creation. The unfolding of life on earth, the coming and going of different species in the Fossil Record, compared with a set of Russian nesting dolls where an encapsulated doll can differ, in preconceived and predetermined ways, from the encapsulating doll, so as to perfectly match the environmental conditions of the geological era into which it was born. Bonnet cited the writings of an Englishman, William Whiston's "*A New Theory of the Earth*" published in 1696[24], and praised by such scientific luminaries as Newton to whom the work was dedicated, when he claimed that the earth underwent a series of catastrophes during its past history, of which the deluge was one, but not the only one. More such catastrophic events were to occur in the future, as was announced in the Holy Bible. Such catastrophic events, marked by the stratigraphic boundaries between the successive layers of sedimentary rock, would wipe out the then existing forms of life. But these harbor in their brain a germ of resurrection, which in turn encapsulates the preformed and preexistent string of generations that would stretch out to the next global catastrophe, which would in turn set free a new set of encapsulated germs of resurrection, which would start a new process of unfolding in a changed world. Following the next global catastrophe, humans might be resurrected as inhabitants of the realm of angels, whereas apes might take the place of worldly humans, according to Bonnet.[25] However, from the beginning of time to the eventual apocalypse of the world, no real change ever took place, nothing new ever developed. All life at any period of earth history was merely the product of growth and unfolding of what already existed. The Creator and nature, according to His plan, work as small as they have to make this possible. To doubt this world-view is to doubt the almightiness of God. Bonnet adopted this doctrine, called ovulism, not only on the basis of his discovery of parthenogenesis, but also primarily as a consequence of the research by Albrecht von Haller, who again had published on the development of chicken in 1758.[14]

Haller was an eminent medical researcher at the University of Göttingen, before returning to his hometown Berne in Switzerland, assuming the role of secretary to the city parliament.[26] Haller had found that the membrane surrounding the yolk in the chicken egg remains in unbroken continuity with the intestinal tissue of the developing chicken throughout its development. The developing chicken must therefore preexist in the egg, not in the sperm, and it had to be preformed in its entirety, with all its parts, for functional reasons. A developing organism cannot be assembled from parts like a machine. If Harvey believed that the heart developed, and hence existed, before its associated vessels, or also before the head and the brain contained within it, he must simply have been wrong. He must have missed parts of the embryo during early stages of its development, because these are perhaps too

---

[24]Whiston, W. 1696. A New Theory of the Earth, From its Original, to the Consumation of all Things. R. Roberts for Benj. Tooke, London.

[25]Bonnet, Ch. 1769. La Palingénésie Philosophique, Vol. 1. C. Philibert and B. Chirol, Geneva.

[26]Balmer, H. 1977. Albrecht von Haller. Paul Haupt, Berne.

small to be seen, or else they are fluid in nature, transparent and hence elusive like the bones. Haller described how bones first appear as soft, transparent, and gelatinous structures, and only successively reveal themselves to observation as more and more "earthy substance" is deposited into them (bones of the internal skeleton are preformed in translucent cartilage before they ossify). All parts of the developing embryo must exist from the beginning of its development, for development means the unfolding of preexisting structures through growth. But above all, growth requires nutrition. The distribution of nutritive material to the unfolding organs requires a transport system, which is the blood-vascular system. However, the blood-vascular system cannot fulfill its role without a pump, which is the heart. In turn, the proper functioning of the heart muscle depends on its proper stimulation through the nerves that innervate it. The nerves, in turn, originate from the brain, and so on. The adoption of the doctrine of preexistence was not only motivated by the desire to avoid the paradox of change but it was also was rooted in empirical observation and in physiological theory construction, which led to the doctrine of the functional correlation of parts in a living organism. On August 11, 1770, Bonnet wrote to his friend Haller: "The animal is evidently a *whole* the parts of which must always have *coexisted*. The heart presupposes the arterial system; the latter presupposes the venous system, etc. . . It suffices to show you a foot, or a hand, for you to be able to fathom the whole. . . The whole universe is a giant machine, of which no part whatsoever could have existed in isolation from all others."[27] The proper functioning of an organism as a whole required the harmonious functioning of all its interdependent parts. The parts of an organism could not be assembled piece-meal during the process of development, nor could those parts be dissociated. Either was believed to disrupt the harmony of the functioning whole and with it the perfect adaptation of the organism to the place in the household of nature to which its species had been assigned by the Creator.

The organism was seen as perfect clockwork, which is designed and constructed according to the blueprint of Creation. This design assured the perfect adaptation of the species to its specific environment. The clockwork cannot function properly if it is assembled piece-meal from its parts until such time as all parts have perfectly been fitted together, and the same would be true for organisms. However, the organism does have to function from the first day of its development. Its parts, therefore, cannot be successively put together by development. All parts of the developing organism must coexist and function together to make development possible in the first place. But then, nothing new ever develops, no change ever takes place. Species consequently cannot change, or so it was believed.

Not just the organism, the entire universe was seen as such a perfectly designed clockwork. The organism was the microcosm that mirrors the macrocosm. The unfolding of the developing embryo mirrors the Great Chain of Being, as both are subject to the same principle of order that puts the complex above the simple. Such was the Plan of Creation, captured by the doctrine of preexistence. Coupled with the

---

[27]Sonntag, 1983, ibid., p. 890.

doctrine of encapsulation, the doctrine of preexistence reduced the change observable in the developing embryo to mere appearance, to the mere actualization or unfolding of what already exists. This was an eighteenth century edifice of French biology that was well aligned with religion and politics, with the ideals of people with wealth and power. Although avoiding the paradox of change, the doctrine of preexistence itself created paradoxes of a different kind, that is, of a biological kind that would ultimately result in its demise. If they were preformed and preexistent in the mother's eggs from the beginning of time, why is it that children not only perpetuate the species *Homo sapiens*, but also beyond that show mixed resemblances to both parents? And why would siblings show different combinations of such parental characteristics? Worse: why would a benevolent Creator preform disadvantaged children handicapped by various malformations before these children were born into the world?

Chapter 3
Changing World Views

Ancient atomistic philosophers explained apparent change in nature by the coming together of indestructible and indivisible, unchanging and eternal atoms in different combinations at different times. The discovery of the seemingly unlimited regenerative powers of the polyp, Hydra viridis, *by Abraham Trembley in 1740 lent empirical support to the view that organisms, their formation and their growth, could be compared to crystals, both being composed of parts. With his theory of embryogenesis, Georges Buffon in 1749 introduced the distinction of organic and inorganic matter. He explained the mixed inheritance of parental characteristics by the offspring with the theory that the embryo forms by the coming together of organic molecules derived from both parent bodies. Maupertuis speculated in 1751 that if accidental 'mistakes' could occur in the coming together of parts during the formation of an embryo, and if these 'mistakes' were, or could become, heritable, a mechanism would be at hand to explain species transformation.*

Nearly 100 years later, Robert Chambers appealed to universal laws of nature to explain change both in the inanimate and in the animated world. The most fundamental laws he thought were the Law of Gravity that ranges over the inanimate world, and the Law of Development, which ranges over the animated world. These laws, or secondary causes, Chambers believed to have been enacted by the Creator, the First Cause, in such a way that both the inanimate and the animate worlds would function as two enormously complex clockworks that run in perfect synchrony. The world would be subject to continuous change, with perfect adaptation being maintained between the ever changing physical world and its inhabitants.

Chambers found the animated world to be subject to an all-pervading principle of order manifest in parallel in the Great Chain of Being, in embryonic development, and in the Fossil Record. This three-fold parallelism he explained through the Law of Development, which pushed nature through a series of transformations that resulted in ever more complex forms of organization, with humans at the top of the ladder of life. Chambers' was an evolving world, but one in which transformation was goal-directed, and hence premeditated by the Creator, who acted through secondary causes, the laws of nature.

O. Rieppel, *Evolutionary Theory and the Creation Controversy*,
DOI 10.1007/978-3-642-14896-5_3, © Springer-Verlag Berlin Heidelberg 2011

3.1   Robert Chambers (1802–1871)

On July 10, 1802, Robert was born as the second son of the Scottish cotton manufacturer James Chambers, in the Village of Peebles.[1] Overwhelmed by the joy and the significance of the event, his parents failed to notice a minor disfigurement, which the newborn child shared with his elder brother William. Each hand and each foot was adorned with six fingers and toes, respectively.[2] Something seemed to have gone wrong during the fetal development of the baby. Some change had occurred, with the addition of an extra digit to hand and feet – a change furthermore that seemed to run in the family. It may be no coincidence that later in his life Robert Chambers would champion an evolutionary world-view. Indeed, as James A. Secord observed,[3] Chambers might have thought of his own body as providing striking evidence for transformation, a sort of transformation he knew that he had inherited. Similar minor disfigurements as those that afflicted Robert Chambers and his brother had been known to occur for a long time. Indeed, the French scientist and philosopher Pierre-Louis de Maupertuis had studied the pattern of inheritance of supernumerary digits over several generations in the family of his friend Jakob Ruhe in Berlin. The surprising conclusion, which he published in his *Système de la Nature* of 1751, was that if organisms suffering from minor malformations could survive, and if these malformations were, or could become, heritable, then species could develop, that is, transform over time: "The multiplication of species would have as its first origin some fortuitous changes. . . each degree of error would have resulted in a new species."[4] Such a conclusion was bound to clash with the received wisdom of the time in many ways. The idea of a new species being born from heritable disfigurements would certainly contradict the doctrine of the intelligent design of organisms, the consequent perfect adaptation of species to their particular place in the household of nature, and the wisdom and benevolence of the Creator that is thereby manifest in the works of nature.

Robert Chambers' life and work would lead him into similar conflicts with the British Establishment that had confronted Maupertuis on the Continent before. He would understand nature as a process of progressive development from simple to more complex forms of life, from mushrooms to humans, along the ladder of life. His *"Vestiges of the Natural History of Creation,"* published anonymously in 1844, presented the Great Chain of Being not merely as a linear hierarchy of static order, a

---

[1]Millhauser, M. 1959. Just Before Darwin. Robert Chambers and Vestiges. Welseyan University Press, Middletown, CT.

[2]De Beer, G., 1969. Introduction. In: G. de Beer (Ed.), Robert Chambers: Vestiges of the Natural History of Creation. University Press, Leicester; and Humanities Press, New York., p. 23.

[3]Secord, J.A., 2000. Victorian Sensation. The Extraordinary Publication, Reception, and Secret of *Vestiges of the Natural History of Creation*. The University of Chicago Press, Chicago, p. 96.

[4]Maupertuis, P.-L.M. 1751. Système de la Nature. In: Oeuvres de Maupertuis, Nouvelle Edition. Corrigée et Augmentée. Nachdruck der Ausgaben Lyon 1768 & Berlin 1758 bei Georg Olms, Hildesheim, 1974, Vol. 2, p. 164.

classification manifest in organismic diversity, paralleled by embryonic development, and anchored in the Plan of Creation. Chambers would come to see the Great
Chain of Being as an order that unfolded through earth history as a consequence of
progressive change. According to him, higher forms of life were not preformed and
preexistent, encapsulated within lower forms of life. Instead, they were the result of
real change, where ancestral forms of life transform into descendant ones that had
not existed before. The term "evolution" would take on an entirely different
meaning: no longer denoting a mere process of unfolding of preexistent structures
of an embryo, it could now denote the development of something new from
something old, the *emergence* of new forms of life. Chambers, however, did not
use the term "evolution" in this sense until he wrote the preface to the 1853 edition
of his book.[5] Rather, he spoke of a "law of development" or a "law of transformation" that placed the power of creation plainly into nature, rather than associating it
with an entity that resides beyond space and time.

It is generally acknowledged that Charles Darwin deserves the credit for having
presented the scientific community as well as the general public, for the first time in
history, with an acceptable scientific theory of evolution. Although this is true,
Darwin was not the first one to put forward a theory of species transformation. His
success hinged on the fact that he was the first one to propose a plausible natural
mechanism to explain the transformation and multiplication of species. The origin
of new species, for Darwin, was an effect of purely natural causes, and in developing his theory, Darwin drew on the work of his predecessors, one of whom was
Chambers: Darwin's "*Origin*" from 1859 can be considered as a work that built on
Chambers' work and the vast and varied responses it triggered.[6] We know that
Darwin heavily annotated his copy of Chambers' book on species transformation,
and it is clear from Darwin's notes that he paid close attention not only to
Chambers' arguments, but also to the criticism leveled against those both by
scientists and theologically motivated opponents to the theory of evolution.[7] He
followed the stunning controversy surrounding "*Vestiges*" to learn how to present
his own argument, and which mistakes to avoid. The controversial reception of
"*Vestiges*" by scientists, clergymen, and the broader public provided Darwin with a
testing ground for his own rhetoric.[8] In that sense, Chambers' book provided *the*
framework for all later discussions of species transformation, including Darwin's.[9]

---

[5]Hodge, M.J.S.1972. The universal gestation of nature, Chambers' *Vestiges* and *Explanations*.
Journal of the History of Biology, 5: 127–151.

[6]Secord, J.A. 1994. Introduction. In: Secord, J.A. (Ed.), Robert Chambers: Vestiges of the Natural
History of Creation, and Other Evolutionary Writings. The University of Chicago Press, Chicago,
p. xliv.

[7]Egerton, F.N., 1970. Refutation and conjecture: Darwin's response to Sedwick's attack on
Chambers. Studies in the History and Philosophy of Science, 1: 176–183.

[8]Secord, 2000, ibid., pp. 431.

[9]Bowler, P.J. 1990. Charles Darwin. The Man and his Influence. Basil Blackwell Ltd., Oxford,
p. 23.

"Vestiges" and Darwin's *"Origin"* came to relate to each other like an ancestor to its descendent: you read the first before the latter.[10]

3.2   Maupertuis' Studies of Patterns of Inheritance

Another author who much earlier influenced the "agenda" for the future development of the theory of species transformation was the already mentioned French scientist and philosopher Pierre-Louis Moreau de Maupertuis (1698–1759),[11] who was elected as the lifetime President of the Academy of Sciences in Berlin in 1746. He was a prolific and progressive author defending the agenda of *materialism* in French biology during the age of Enlightenment. Quite generally, materialism seeks an explanation of natural processes without taking recourse to transcendental powers or entities that reside beyond the realms of matter, time, and space. As mentioned earlier, some members of the Ruhe family he befriended in Berlin had had six digits in hands and feet. A supernumerary digit, another "part" seems to have been added, by accident, to the hands and feet of these children. To Maupertuis, this only proved an idea that he and others had before, which was that embryos formed by the juxtaposition of parts as they develop, in analogy to the growth of crystals. These initially minuscule parts were thought to be derived from the seminal fluids of both father and mother, which came together and were mixed up with one another in the female uterus upon conception. The subject of the existence of a female seminal fluid remained a matter of debate through most of the eighteenth century and beyond, in spite of the fact that the eminent medical researcher Albrecht von Haller had conclusively argued against its existence as early as 1752: "I find nothing that could convince me that the fair sex produces a semen, let alone that it would release a seminal fluid, which would then mix with the male fluid."[12] The issue of a female seminal fluid is a good example of scientists invoking the existence a theoretical, that is, unobservable substance, such as the female seminal fluid, to explain observable phenomena, which in this case is the inheritance of a variable mixture of maternal and paternal traits by offspring. A materialistic philosophy as the one espoused by Maupertuis held that the initial juxtaposition of parts in the formation of the embryo was not guided by some mysterious agent that resided somehow outside or beyond matter, such as an immortal soul bestowed upon an organism by the Creator. Instead, the coming together of parts in the formation, that is, in the development of an organism was contingent upon properties and dispositions entirely inherent in the minuscule material particles that were to form the organismic "whole." Consequently,

---

[10]Secord, 2000, ibid., p. 39.

[11]Millhauser, 1959, ibid., p. 63.

[12]Haller, A.v. 1752. Vorrede über des Herrn von Buffon Lehre von der Erzeugung, p 105. In: Sammlung kleiner Hallerscher Schriften, 2nd Ed., 1772. Emanuel Haller, Berne.

accidental mistakes could happen in the initial juxtaposition of parts, such as the occurrence of supernumerary digits.

How is it that a child resembles both father and mother, and has characteristics of its own as well? How is it that brothers and sisters who are not twins are not identical, but instead show different and variable patterns of inheritance of individual maternal and paternal characteristics? But then, how can there be the occasional twins? The theory of a preexisting embryo, encapsulated either in the male spermatozoon or in the female egg, could not explain such variable inheritance of parental characteristics by the offspring, let alone any possible disfigurements, nor could it explain the origin of twins. Charles Bonnet's feeble attempts to explain away such weaknesses of his favorite doctrine (sketched in the previous chapter) through differential pressures that were accidentally exerted on the "unfolding" embryo during gestation were clearly ad hoc and unconvincing. Formation of the embryo by the juxtaposition of parts or particles provided by seminal fluids from both mother and father was how Maupertuis and other authors of his time explained the phenomena of malformation, indeed the phenomena of reproduction and heritability, in general. Driven by natural curiosity, Jakob Ruhe had carefully mapped the occurrence of supernumerary digits on his family tree, and to his surprise detected some regularity. Maupertuis established himself as a pioneer in genetics[13] when he interpreted this regularity as evidence for mechanisms of inheritance: "One can see from this genealogy, which I have followed with exactitude, that sexdigitism is transmitted equally by the father and the mother."[14] From his studies he concluded to a most heretical perspective. Suppose, as Maupertuis did, that accidental variations (such as the addition of supernumerary fingers and toes to hand and feet) were possible and heritable, as he found to be the case in the Ruhe family, would this not provide the starting point for the origin of new species? But if such accidental changes could add up to the transformation of a species, then the species could have no essential properties. The species potentially becomes subject to real change through time, a change that is not just the "becoming visible" of what already existed from the beginning, but the emergence of something new. Without essential, unchanging properties, species were free to transform: "I am quite willing to accept that these supernumerary digits are by virtue of their first origination nothing but accidental varieties... but once they have been established through a sufficient number of generations of which both sexes have been afflicted, then these varieties give rise to species. It is, perhaps, in this way that all species have multiplied."[15]

Authors like Charles Bonnet and Albrecht von Haller, who defended the doctrine of the encapsulation of preformed and preexisting embryos, decried the theories

---

[13]Glass, B. 1959. Maupertuis, pioneer of genetics and evolution, pp. 84–112. In: Glass, B., O. Temkin and W.K. Straus Jr. (Eds.), Forerunners of Darwin 1745–1859. The Johns Hopkins University Press, Baltimore.

[14]Maupertuis, P.-L.M. 1753. Les Oeuvres de Mr. de Maupertuis, Vol 2. E. de Bourdaux, Berlin, p. 386.

[15]Maupertuis, 1753, ibid., p. 387.

proposed by Maupertuis and like-minded biologists not only as scientifically wrong, but also as ethically objectionable, since they threatened to lend (in their view unjustified) scientific authority to atheism.[16] To recognize this threat, one only needs to return to Ancient Greek philosophers again, but this time to thinkers such as Heraclites and Epicurus, who laid the classical foundation for a materialist atomistic philosophy. Ironically, the motivation for them to develop their atomistic theories was the same that led Bonnet and Haller to adopt the doctrine of preexistence: it was to avoid the paradoxical problem of change. According to the Ancient Greek atomists, all material objects are composed of minuscule particles, called the atoms. These atoms were the indivisible, indestructible, eternal, and never changing building blocks of nature. Any apparent change in nature was parasitic on the coming together of eternal, never changing atoms. Atomists explained the differences between natural objects and their change through time, as an effect of the coming together of different atoms in different combinations at different times. For atomists, the coming together of these atoms in different combinations to form the changing objects we observe in nature was not guided by a soul, however, or any other mysterious force that transcended matter, space, and time. There was no need for a guiding principle of development and change, since the capacity of coming together in the right way to form a fully functional living being was inherent in the atoms, that is, in matter itself. This is where Bonnet and von Haller located the threat of atheism. But, as we know from the previous chapter, they rejected an atomistic conception of organisms also on what they accepted as scientific, that is, empirical grounds. In their view, the "clockwork-model" of organisms and the consequent doctrine of the functional correlation of their parts did not allow these to develop by the piece-meal aggregation of parts. According to them, life seemed to require the harmonious cooperation of all the parts of the integrated whole, which is the complex organism. The resurrection of an atomistic conception of organisms by Maupertuis and his fellow revolutionaries, therefore, required some very strong arguments indeed, and these were provided by Abraham Trembley's discoveries that triggered a scientific revolution.

3.3   Trembley's Experiments with the Freshwater Polyp

Materialistic ideas and theories such as those of Maupertuis and his contemporaries were sort of "in the air" at the time. In fact, they almost imposed themselves on the inquiring mind following the "discovery" of the green polyp, *Hydra viridis*, by Abraham Trembley in 1740.[17] Abraham Trembley was none other than the cousin

---

[16]Sonntag, O. 1983. The Correspondence between Albrecht von Haller and Charles Bonnet. Huber, Berne.

[17]Trembley A. 1744. Mémoires pour servir a l'Histoire d'un Genre de Polypes d'Eau Douce. Durand, Paris. For details and further references see Rieppel, O. 1988. Fundamentals of Comparative Biology. Birkhäuser Verlag, Basel.

of Charles Bonnet. He served as a house-teacher at Count Bentnick's estate in The Netherlands. One day he took the two boys he tutored to a nearby brook to collect water plants for study. As he contemplated the duckweed he had brought back in his powder jar, he spotted a strange organism that he had never seen before. Ignorant of Antony van Leeuwenhoek's earlier description of the organism, Trembley thought he had discovered a new species, which Linnaeus in 1767 would name *Hydra viridis*. The genus *Hydra* includes soft-bodied animals whose very simple anatomy resembles an upright standing bag with a series of movable tentacles lining its opening. With these tentacles the animal catches minute prey items, which it then forces into the body cavity to be digested. The creature that Trembley had collected gets its green color through the peculiar fact that symbiotic green algae (the unicellular alga *Chlorella*) live inside the cells that line the body cavity of the polyp, a fact which Trembley and his contemporaries had no chance to discover. Although Robert Hooke described the cellular nature of plants when examining the microscopic structure of cork as early as 1665,[18] the recognition of cells as the fundamental unit of life, capable of replication and differentiation, had to await the publication of the research of Theodor Schwann and Matthias Jakob Schleiden (in 1839), and of Rudolf Virchow (in 1858). At the time of Trembley's investigations, there neither existed a mature cell theory, nor was there any evidence known that would indicate even only the faintest possibility of a plant living inside an animal, so to speak. All that Trembley recognized were these fragile living beings of green color that attached themselves to the substrate and carried a ring of "filaments" that were freely floating about. At first, he identified the creatures as a kind of water-plant, their tentacles as floating roots. Their green color and the way of multiplying by budding he certainly recognized as classical characters typical of plants, phenomena that were believed to be governed by a "vegetative soul." However, continued observation revealed voluntary movements of these creatures, which performed somersaults to move from one place to another. He also observed the strange organisms to actively catch minute prey with their tentacles, and when stitching them with a needle, he observed them to quickly contract, as if they felt some sort of pain. To move about and to actively catch prey are the characteristics classically considered to be typical of animals. In addition, the eminent Albrecht von Haller had identified the sign of "*irritabilité*," the contraction of an organism upon being stitched with a needle, as a diagnostic feature of "animal fibers," believed to be governed by an "animal soul." Were those creatures plants or animals? Trembley decided to tackle the problem experimentally.

At that time, it was well known that a whole plant could regenerate from a single twig, even from a single leaf cut from a tree or any other plant. Animals do not usually show the same powers of regeneration. If a feeler is cut off from a snail, if a limb is cut off from a salamander, or if a lizard lets go of its tail in a defense reaction, the animals will regenerate the missing parts, but the severed parts will not

---

[18]Rudwick, M.J.S. 1972. The Meaning of Fossils. Episodes in the History of Paleontology. Macdonald, London, pp. 54f.

regenerate in to new whole animal. In fact, Aristotle had already used this difference in regenerative power to distinguish plants from animals. To find out whether he was dealing with a plant or an animal, Trembley cut *Hydra* into two, four, eight, and more pieces. He was surprised to find that every single part, as many as there were, regenerated to form a new whole organism. He grafted the ring of tentacles of one animal on to another, or he combined parts of different individuals in the formation of a new whole organism. Cutting two polyps horizontally, he could induce the foot-part of one individual to fuse with the tentacled top-part of the other individual in the formation of a new "mixed" organic whole. Obviously, the first conclusion he drew from these observations was that animals are composed of "parts." Second, those "parts" could be recombined and exchanged in the formation of a functioning "whole." Third, the animal nature of the organisms was expressed by the contraction, the "*irritabilite*" of its fibers upon stimulation.

At the time of Trembley's experiments, in the middle of the eighteenth century, such "animality," that is, the capacity of movement, reaction to external stimulation, was believed to be imparted to the body by the (animal) soul. Body and soul were, and by some still are, considered as two fundamentally different substances, one material (body), the other immaterial (soul). While it was easily understood that material bodies could be chopped up, the immaterial soul seemed to lend itself less easily to being cut into pieces. Trembley cut a specimen of *Hydra* horizontally into two parts. He considered the part that attaches to the substrate to be the foot part. The head part then is the part that carries the ring of tentacles around its opening. At the time, the brain was considered to harbor the human soul; by analogy, the polyp's soul that imparted on the creature the capacity of movement and "irritability" would have to be located in its head portion. As the tentacled head part of a polyp regenerates a new foot portion, there seemed to be no problem to explain how it could be that the new "whole" organism would show *irritabilite*. This was easily explained by the fact that the new organism was governed by the soul located in the head part from which the process of regeneration had started. Not so in the case of the "whole" new organism that regenerated from the severed foot portion of a polyp. It, too, showed *irritabilite*, but where did it get its soul from? "Could there be souls that can be cut in half?" was the question raised by the famous entomologist René de Réaumur,[19] who immediately set out to repeat Trembley's experiments once he was informed of them by letter in 1741,[20] only to find the latter's observations fully confirmed. The conclusion, immanent in all these experiments, but not spelled out by Trembley himself, was that either there is no soul in animals, or that the animal soul is coextensive with matter, that is, it could not be distinguished from matter.

[19]Réaumur, R.A.F. de. 1742. Mémoires pour Servir à l'Histoire des Insectes, Vol. 6. Imprimérie Royale, Paris, p. lxvii.
[20]Trembley, A. 1943. Correspondence inédite entre Réaumur et Abraham Trembley. Introduction by Emile Guyénot. Georg & Cie. Geneva.

It was the enlightened philosophers of the time who seized upon such conclusions. Julien Offray de LaMettrie[21] was a French physician and philosopher living in Berlin, where Maupertuis also lived. Thus removed from the reach of the Church officials of Paris, he published his then "scandalous" book "*L' Homme Machine*" ("Man a Machine") in 1747. In the course of his prose he exclaimed: "Look for yourself at Trembley's polyp! Does it not incorporate within itself the power of regeneration? How absurd would it therefore seem to be to believe that causes exist to which end everything has been created... causes the ignorance of which we will never be able to conquer and which hence leads us to believe in a God who, according to some, is not even a rational being. To destroy in this way the power of chance is not yet to prove the existence of God, since there might be something which is neither pure chance, nor Divine Being, but nature..."[22] Its polemic style aside, the crucial point in this quote is the denunciation of a belief in causes *to which end everything has been created*. This belief embodies an appeal to purposefulness in nature, to a goal-directedness of natural processes such as embryonic development, and ultimately to the handwriting of an ordering and planning intelligence that is believed to be manifest in the works of nature. This is the conclusion which LaMettrie rejected with reference to Trembley's polyp, thus echoing one of the intellectual leaders of the French Enlightenment, the editor-in-chief of the famous "*Encyclopédie*," which set the revolutionary tone of the time – Denis Diderot: "If nature presents herself in the guise of a Gordian knot, we should not aspire to cut it with the help of the hand of somebody who will, eventually, reveal himself as a knot even more difficult to cut."[23] This, of course, is reminiscent of the warning issued by the Roman poet Horace to those authors who call for Gods to interfere with the natural course of events: "*Nec Deus intersit, nisi dignus vindice nodus*" ("do not let God interfere if the knot can be untangled without Him").[24]

With Abraham Trembley's experiments and their manifold confirmation by other authors, materialism had found an empirical, that is, observational base in biology. The power of regeneration was found to be an utterly intrinsic property of the polyp's tissues, a property not powered by an immaterial soul occupying a realm beyond matter. Of course, this did not prevent naturalists of later generations to variably ascribe to plant and animal tissues some esoteric vital forces that would be immanent in nature. They did so due to the lack of a better explanation of observed natural processes that were demystified by modern science only. Chambers was one of them, as he invoked some mysterious force in his explanation for the evolution of

---

[21]Millhauser, 1959, ibid., p. 64.

[22]La Mettrie, J.O. 1865. L'Homme Machine. Avec une Introduction et des Notes de J. Assézat. Frédéric Henry, Paris, p. 102–103.

[23]Diderot, D., 1749 [1972]. Lettre sur les Aveugles. Garnier-Flammarion, Paris, p. 103.

[24]Similarly invoked with respect to the Darwin-Wallace theory of evolution; see 'Wallace defends Darwin's priority – 50 years on', p. 46 in: Survival of the Fittest. Celebrating the 150th anniversary of the Darwin-Wallace theory of evolution. The Linnean Special Issue No. 9, The Linnean Society, London.

life on earth that would push organisms up the Ladder of Life. But others entered that stage before him.

3.4   Georges Buffon's Evolving World

Georges-Louis Leclerc, Comte de Buffon (1707–1788),[25] Director of the Royal *Jardin des Plantes* in Paris, again immediately recognized the significance of Trembley's findings. In the first volume of his *Histoire Naturelle, Générale et Particulière*, a multi-volume treatise on natural history that was published from 1749 through 1788, Buffon expounded a theory of embryogenesis that compared animal growth to the growth of crystals: the embryo, just as crystals, is formed by the juxtaposition of elementary particles, as Buffon thought to have been demonstrated by Trembley's polyp. But to distinguish between the crystal and the living organism, Buffon drew the fundamental distinction of organic vs. inorganic matter. Borrowing from the Ancient atomists, he believed organic matter to be composed of indestructible, indivisible particles, the fundamental building blocks of all living matter, which he called "*molécules organiques.*" These organic molecules would compose the topsoil, from where they were taken up by the plants through their root system. The organic molecules would further enter the animal food chain through herbivores. Upon death and decay of plants and animals, the organic molecules would return to the topsoil. The result is again a world of dynamic permanence, where indestructible organic molecules that undergo no essential change cycle endlessly from the soil through plants and animals back into the soil. Coming from the soil and returning to it meant that these organic molecules did not undergo any essential change: they remain essentially the same indestructible building blocks of life throughout their different metamorphoses, which they undergo as they cycle through plants and animals. Plants take up organic molecules from the soil in which they grow, animals obtain them by eating plants, and carnivores obtain them by eating herbivores. The development and growth of plants and animals was fuelled by the assimilation of these organic molecules into their tissues. Through their assimilation into the growing body, the organic molecules would take on the properties of the specific living tissues that they became part of, but such change could only be an accidental change. In the fully-grown organism, a surplus of organic molecules would be stored in the reproductive organs. Conception would bring seminal fluids from both parents together in the female uterus, where the organic molecules would be mixed before they are recombined to form the embryo. Buffon once again invoked a female seminal fluid, which he believed to be generated by the *corpus luteum*, to which Haller replied: "It is absolutely certain

[25]Rieppel, O. 2001. Georges-Louis Leclerc, Comte de Buffon (1707–1788), pp. 31–50. In: Jahn, J., and M. Schmitt (Eds.), Darwin & Co., Vol. 1. C.H. Beck, Munich.

that this yellow gland is the consequence, not the cause, of impregnation."[26] Buffon's theory was designed to explain the phenomena that were left unexplained by the doctrine of preexistence, which are the mixed inheritance of maternal and paternal characteristics by the offspring, as well as the occasional malformations. In contrast to the doctrine of preexistence, however, Buffon's theory is less successful in explaining design, purpose, and goal-directedness in nature, from which resulted the perfect adaptation of the species to which the developing embryo belongs.

Animal generation, for Bonnet as much as for Buffon, served the perpetuation of the species, as much as of its perfect adaptation to its specific environment. Bonnet's account, detailed in the last chapter, is a theory of Russian nesting dolls, which explains the perfect perpetuation of species in an almost trivially easy way. But what are the exact mechanisms that make this possible on Buffon's account? According to Buffon, the organic molecules first and foremost become assimilated to the organism which they entered in support of the growth and metabolism of its tissues. When an animal has grown to its adult size, organic molecules would still be required to sustain metabolic processes, but since growth had come to an end (or had dramatically slowed down), there would be a surplus of organic molecules taken up with the food. These superfluous organic molecules would circulate through the parent body in its blood stream on their way to storage in the reproductive organs. During such circulation, a mysterious force, called an "internal mold" by Buffon, would impart on these organic molecules individual accidental properties of the parent body, as well as the essential properties of the species to which the parent organism belongs. As the seminal fluids of both parent organisms, father and mother, would come together in the female uterus upon conception, another mysterious force, which Buffon called a "formative force," would draw the organic molecules together in variable combination to form the embryo, an integrated whole. That way, the offspring inherited not only a mixture of both maternal and paternal characteristics, but also the essential properties of the species to which the parent organisms belonged. These essential species-specific properties would be essential properties for the developing embryo, preserving its perfect adaptation to the world into which it was going to be born, but they would not be essential properties of the organic molecules, which would eventually return into the topsoil.

Buffon's theory dispensed with the immaterial soul as the essential property of the developing organism, but instead invoked mysterious "internal molds" and a "formative force" to explain the persistence of the species-specific form in the developing organism, and consequently the maintenance of design and purpose in the coming together of parts in order to preserve the goal of the functional integration and perfect adaptation of the organisms. The only problem remaining was the functional correlation of the parts: the blood-vascular system could not function without the heart, the heart could not function without innervation, the nerves could not function without a brain, and so on. Buffon solved this problem by claiming that

---

[26]Haller, 1752, ibid., p. 107.

the juxtaposition of the molecules that form the embryo would happen instanta-
neously, in the duration of an "inkling of the eye."

Buffon felt confident that his theory of embryo formation would explain not only
patterns of inheritance, but also the possibility of malformation, such as the addition
of an extra digit to hands and feet. Buffon's theory materialized the essence of the
ever-changing organism in terms of properties that are inherent in matter, imprinted
on the organic molecules. The "internal mold" and the "formative force" were
mysterious in the sense that their mechanistic basis eluded Buffon and the science
of his time. They were not meant to be entities comparable to a soul, but understood
as a nonmaterial substance (principle) that is fundamentally different from matter.
Buffon's mysterious forces are fundamentally inherent in matter, although it eluded
the science of his time how matter could provide a substrate for their functioning.
Buffon saw nothing wrong with such a postulate, as the same was true for Newton's
celebrated gravitational force. But it is important to recognize how Buffon avoided
the paradoxical problem of change by drawing the distinction between accidental
and essential properties. The properties that the organic molecules took up from the
parental bodies were accidental properties, transmitted to the offspring in a variable
pattern.

This was real change, since the offspring would combine characteristics of both
mother and father in a novel combination. The accidental addition, or deletion, of
parts, with more or less serious consequences of malformation, likewise documented real change. But at the bottom line, the organic molecules would take on the
species-specific and the individual parental characteristics only transiently. At the
end of the cycle, the organic molecules return to the topsoil, to the condition from
which they started their ascent into organic beings. What appears to be real change
in organisms from conception through growth and maturation to senescence and
death is parasitic upon the coming together of essentially unchanging parts in
different combinations and adorned with different changing properties at different
times. An ambiguity remained, however. To call the individually variable properties of parents and offspring accidental ones was easily acceptable. But the organic
molecules would also transfer the species-specific properties of the parents to their
offspring, and these properties would be considered essential properties of the
species to which the parents and their offspring belong. Since the organic molecules
did not themselves undergo any essential change, the conclusion must be that what
for them could only be accidental properties were nevertheless essential properties
of the species to the replication of which they contributed. Perhaps, it is for this
reason that throughout his *oeuvre*, Buffon displays a much-discussed ambiguity with
respect to the question of species transformation, which, if it happened, would rob
the species of its essential properties.[27] At the end of the day, however, he settled
on the conclusion that species are "perduring entities, as ancient and permanent as

---

[27]Bowler, P.J. 1973. Bonnet and Buffon: theories of generation and the problem of species. Journal
of the History of Biology, 6: 259–281. Farber, P.L. 1972. Buffon and the concept of species.
Journal of the History of Biology, 5: 259–284. Roger, J. 1971. Les Sciences de la Vie dans la
Pensée Française du XVIIIe Siècle, 2nd Ed. Armand Collin, Paris.

nature itself,"[28] in spite of the fact that he allowed limited adaptive change (which he called "*dégénération*") within species.

The world he sketched of ever changing combinations of essentially immutable and indestructible organic parts, Buffon embedded in a cosmology that incorporated the passing of nature, the arrow of time. Buffon adopted the "cooling earth theory" to explain physical changes on the surface of the earth over geological time. As it spun off from the sun, the earth formed a fiery ball, which since then had continuously cooled off. The Polar Regions were the first to become inhabitable by living organisms, which still had to be adapted to a cold environment, however, as the wooly mammoth doubtlessly was. The first well-preserved mammoth mummy had, indeed, been dug up from the permafrost soil of Siberia by the time of Buffon's writing.[29] In contrast, the equatorial regions were too hot to be colonized during early phases of earth history. Further cooling of the earth rendered equatorial regions gradually habitable, while it was becoming concomitantly increasingly colder at the Polar Regions. As a consequence, the mammoth populations tended to push southwards, eventually adapting to the new climate by loosing their thick fur and becoming somewhat smaller. But how was such a change, a process called "degeneration" by Buffon, possible for a species that was to be marked out by essential, never-changing properties? Indeed, and in contrast to some other famous places in his monumental book series on natural history, Buffon maintained the constancy of species in his discussion of the "degeneration" of the mammoth to the elephant.

Because of the cooling of the earth, the organic molecules that circulate endlessly from topsoil through plants into animals and back into the soil are exposed to a gradually changing physical environment. The internal molds of reproducing organisms together with the formative force would orchestrate the coming-together of the organic molecules in such a way that the essence of the species and its perfect adaptation to its specific environment would be perpetuated. However, on Buffon's cooling earth theory, the environment undergoes continuous change, such that the maintenance of perfect adaptation required a potential for adaptive change of the species. Buffon was thus forced to combine his essentialistic view of species with a theory of adaptive change, for which he believed to have found a striking example in the "degeneration" of mammoths of past times that have become elephants living today. He explained such adaptive changes as a consequence of changing qualities of the organic molecules. Cycling through the soil following the death and decay of plants or animals, the organic molecules are exposed to the changing climate, and as a consequence they themselves undergo change that would reflect the drop of temperature. Therefore, in addition to taking up properties from the parent bodies through which they circulate, the organic molecules would also be able to pick up properties that reflect the ever-changing environment. Becoming assimilated to the

---

[28]Lovejoy, A.O. 1959. Buffon and the problem of species, pp. 84–113. In: Glass, B., O. Temkin and W.K. Strauss jr. (Eds.), Forerunners of Darwin 1745–1859. The Johns Hopkins Press, Baltimore, MD. The quote is from p. 101.

[29]Rudwick, M.J.S. 2005. Bursting the Limits of Time. The Reconstruction of Geohistory in the Age of Revolution. The University of Chicago Press, Chicago, p. 265.

developing and growing animal, the latter's constitution would likewise become adapted to the changing environment. But again, wavering back and forth on the issue of species transmutation, Buffon eventually found peace in the conclusion that such adaptation could never generate entirely new types of animal species. This is, perhaps, why he saw such adaptational processes not only as processes of species transformation but also as processes of species "degeneration," that is, as processes through which a species departs from the initial adaptation to the place in the household of nature to which it had first been assigned by its Creator. Nature for Buffon continued to be partitioned into distinct species that remain separate, identifiable, and reidentifiable through time and space. According to Buffon, mammoth and elephants belong to the same species. Species do not undergo essential change, but species boundaries might become a little blurred due to environmental influences, as was demonstrated by the mammoth and the elephants, or by the horse and the donkey. But such blurring of species boundaries was a problem only for the inquiring naturalist, who seeks to distinguish, identify, and reidentify animal species on the basis of their external characteristics. It was not a problem located in nature, which, for Buffon, remained carved up into discrete types or species.

To be sure: Maupertuis and Buffon, like Darwin a little more than a hundred years later, did not have a valid theory of inheritance. In his 1859 "*Origin*," Darwin recognized the importance of individual variation for evolution to occur, but while he talked about the inheritance of parental traits by offspring, or of ancestral traits by descendants, he had no mechanism to offer as an explanation. However, later in his life, Darwin did develop a theory of animal embryogenesis, which also explained the phenomena of mixed inheritance of parental features, a theory in which atomic building blocks of life which Darwin called "gemmulae" played much the same role as did the "*molécules organiques*" in Buffon's theory.[30] In a letter from the year 1865, Darwin announced to his friend Thomas Huxley: "I have read Buffon: whole pages are laughably like mine. It is surprising how candid it makes one to see one's views in another man's words. I am rather ashamed of this whole affair..."[31]

3.5 Chamber's "Vestiges": An Evolutionary World View

Likewise ignorant of the mechanisms of inheritance, Robert Chambers picked up the question of species transformation raised by Maupertuis just over 50 years before he was born, yet pursued it in a much more general sense. First, however, the

---

[30]Sloan, P.R. Darwin's invertebrate program, 1826–1836, pp. 71–120. In: Kohn, D. (Ed.), The Darwinian Heritage. Princeton University Press, Princeton. Hodge, M.J.S. 1985. Darwin as a lifelong generation theorist. pp. 207–243. In: Kohn, D. (Ed.), The Darwinian Heritage. Princeton University Press, Princeton.

[31]Darwin Fr. 1887. The Life and Letters of Charles Darwin, 3rd ed. John Murray, London.

young man tried to make a living as a teacher, a clerk, and finally as a bookseller before, in 1832, he joined his brother in the successful Chambers publishing company in Edinburgh. Robert was a bookworm! Slowly recovering from surgery performed on his hands and feet to remove the supernumerary digits, he used his time to build the foundation of his self-taught expertise in sciences that ranged from cosmology through geology to biology. He devoured every work on natural history he could get a hold of, and in that way obtained knowledge of the ideas and hypotheses of his predecessors such as the ones just discussed. As noted by James A. Secord, it was "atheists and Frenchmen"[32] who discussed species transformation – French being a language which Chambers was comfortably fluid in reading[33] given the "classical education"[34] that he had received in his school days.[35] Geology was his favorite subject, and through his later publications in that subject area, especially on sea level changes and glaciation, he even earned somewhat of a scientific reputation in this field.[36] But the remains of once living organisms, the fossils hidden between layers of rock that formed the earth's crust, likewise caught his interest, and spurred his imagination. In the field of biology and paleontology (the study of fossils), Robert Chambers could never lay claim to any more respectable position than one which his opponents, the contemporary scientific community, dismissed as that of a dilettante amateur. Nevertheless, Chambers was instrumental in gaining public acceptance for the theory of glacial ages put forward by the contemporary geologist and paleontologist Louis Agassiz.

The latter scientist[37] was a native Swiss, who after studying at various universities in Switzerland and Germany was appointed as a professor of natural history at the Lyceum of Neuchâtel in the French speaking part of Switzerland in 1832. There he pursued his investigations of living and fossil fishes, but also took an interest in glaciology. He studied the structure and movement of the ice in the glaciers of the Swiss Alps and its influence on the surrounding geology, which culminated in his "Etudes sur les Glaciers" of 1840. In the course of this work, Agassiz became aware that the grinding action of a glacier left well-defined geological traces in areas from which the ice later retreated. Using these geological clues, Agassiz inferred that past earth history had witnessed a veritable ice age, large areas of the continents having been covered by massive and far-reaching sheets of ice. In 1846, Agassiz left his family and country behind owing to debts he had incurred in the private publication of his studies on fossil fishes in the format of magnificently illustrated monographs that were published from 1833 through 1843. Agassiz

---

[32]Secord, 2000, ibid., p. 92.

[33]Millhauser, 1959, ibid., p. 83.

[34]Secord, 1994, ibid., p. xviii.

[35]For an in-depth historical analysis of Chambers' sources, and the degree to which his knowledge of Continental writings was based on secondary literature, see Secord, 2000, ibid.

[36]Secord, 2000, ibid., p. 391.

[37]Winsor, M.P. 1991. Reading the Shape of Nature. Comparative Zoology at the Agassiz Museum. The University of Chicago Press, Chicago.

continued to teach and pursue his research interests in Boston.[38] In 1848, he was appointed as professor at Harvard University and in that capacity he founded the Museum of Comparative Zoology of Harvard University in Cambridge[39] in 1859, the same year that saw the publication of Darwin's *"On the Origin of Species."* A leading paleontologist and zoologist of his time, he was to become a leading opponent of Darwin's theory of evolution. But long before these tribulations were to appear in the scientific literature, the interaction with Agassiz' and other geologists' work left Chambers with a sense of deep time, with an appreciation of repeated climate changes throughout earth history, and quite generally with a sense of change as one looks at the fossil fishes recovered from successive layers of the rocks exposed in northern Scotland and elsewhere.

In 1841, Robert withdrew from Edinburgh to settle in St. Andrews, where he hoped to find the time and leisure to complete his major project. The plan was to distill, from the scientific literature of his time, the evidence supporting his grandiose vision of an ever-changing world. He was perfectly conscious of the risks he would incur with such a project, which is why he took every conceivable precaution to remain anonymous. The manuscript, at that time to be submitted to the publisher in handwriting, was drafted by his wife Anne, and a friend living in Manchester, Alexander Ireland, had to act as an intermediary between the author and his publisher in London.[40] The reasons for such elaborate precautions were manifold, but perhaps predominant amongst them was the fear that if the author, who had received no formal training in natural sciences, were revealed, his work could all too easily be dismissed as amateurish by the scientific community – as indeed it was. By remaining anonymous, Chambers had hoped to earn the serious close attention of contemporary scientists that he thought his synthesis deserved. His work indeed received serious and close scrutiny by professional scientists, but it resulted in an avalanche of forceful criticisms of scientific detail that caught Chambers completely off-guard.[41] The leading scientists of Victorian England were upper-class gentlemen, mostly independently wealthy, who claimed expertise and authority in a certain field of inquiry they called their own, while taking care not to step on the toes of fellow researches. Boundaries were drawn not only between what is or should be science and what is not, but also within science, demarcating one department from the other: it was declared impossible for anyone to claim well-grounded expertise across a broad range of scientific disciplines.[42] Chambers efforts flew in the face of convention, as he tore down those boundaries, integrating

[38]Lurie, E. 1960. Louis Agassiz: A Life in Science. The University of Chicago Press, Chicago.

[39]Winsor, 1991, ibid.

[40]Secord, 1994, ibid., p. xxxix.

[41]Secord, 1994, ibid. xxxiv.

[42]Secord, 2000, ibid., p. 244.

the sciences from different departments with one another and weaving them together in support of a comprehensive view of an evolving universe.[43]

In October 1844, "*Vestiges of the Natural History of Creation*" appeared on the market: the book triggered a heated controversy that eclipsed all other ongoing debates about contemporary scientific and philosophical issues.[44] It provoked uproar among the religious and scientific establishment of Victorian England, but the avalanche of insult, discredit, and ridicule could not dampen the sales of the book. Opposed by the scientific establishment, Chambers' treatise enjoyed admiration, praise, and even consent amongst the broader public.[45] Chambers answered his critics with successive new editions of his book. By 1860, the treatise had gone through 11 editions, sold over 20,000 copies, and had been exported to the Unites States, Germany, and the Netherlands.[46] However, the harder Chambers tried to defend his cause by the inclusion of new material in his text, the more he disclosed his lack of expertise in biological matters. But the bottom line was that the Scotsman fought a battle against the Victorian Establishment, which in some loose sense paved the way for Darwin.

Chambers basic thesis was as simple as it was naïve: The world of organic beings is subject to an all-pervading "Law of Development" in the same way as the world of inorganic matter is governed by the all-pervading "Law of Gravity." Following the leading scientists and philosophers of his time, Chambers found the natural course of events to be governed throughout the universe by universal laws of nature. He argued that there were two such laws of highest order and greatest generality: the law of development, governing the animate world, and the law of gravity, governing the inanimate world. His background led Chambers to believe that the world we live in is one of continuous change governed by these two great laws of nature. He could no longer believe that the world, with all its parts and diversity, should have been created "at the beginning," and should never have undergone any change since. Indeed, at the time of his writing, Geology had clearly spoken out against the idea of an initial Creation, taking place within as little time as six days. Evidence had amassed which would allow an allegorical interpretation of the Book of Genesis only: geological epochs were allegorically equated with "days" of Creation – the tradition of "scriptural geology"[47] was borne.

Chambers found support for his views in the writings of the foremost astronomer of his time, Sir John Frederick William Herschel, who catalogued "nebulae" or what looked like clouds of uncondensed gas hovering in space, following his

---

[43]Barnes, B., D. Bloor, and J. Henry. 1996. Scientific Knowledge. A Sociological Analysis. The University of Chicago Press, Chicago, p. 157.

[44]Secord, 2000, ibid., p. 1.

[45]Yeo, R. 1984. Science and intellectual authority in mid-nineteenth-century Britain: Robert Chambers and *Vestiges of the Natural History of Creation*. *Victorian Studies*, 28: 5–31.

[46]Williams, W.C. 1971. Chambers, Robert, p. 191–193. In: Gillispie, C.C. (Ed.), Dictionary of Scientific Biography, Vol. III. Charles Scribner's Sons, New York.

[47]Millhauser, 1959, ibid., p. 55.

father's discovery of a new "nebulous star" that later turned out to be a planet.[48] As understood by Chambers: "We have seen reason to conclude that the primary condition of matter was that of a diffused mass, in which the component molecules were probably kept apart through the efficacy of heat; that portions of this agglomerated into suns, which threw off planets; that these planets were at first very much diffused, but gradually contracted by cooling to their present dimensions."[49] According to the nebular hypothesis, chaos would have ruled amorphous matter at the beginning of the universe, but through the continuous action of laws imposed on nature, this chaos was slowly but steadily transformed into an orderly planetary system subject to uniform motion. The notion of *natural laws* pervades Chambers' writings as a trademark: "It is remarkable of physical laws, that we see them operating on every kind of scale as to magnitude, with the same regularity and perseverance."[50] The Universe was not put into place at once, at the beginning of time. Instead, the universe, as we know it came into being, developed or evolved from original chaos that came under the rule of the Law of Gravity. Chambers found this law, like any other universal law of nature, to operate with the same regularity and perseverance throughout time and space. The universe came into being through the action of these laws, but once it had formed, it functioned like a clockwork. It is the regularity of the movements of the heavenly bodies that suggested an underlying lawfulness. Once inferred from the regularity of naturally occurring events and processes as revealed by observation, Chambers, like the scientific authorities of his time, then turned the argument around in using the established law inferred from observed regularity as an explanation for the coming into being of this regularity.

Chambers found further evidence for change in the physical world engraved in stone. Why did quarries, mountain slopes, and coastlines exhibit a stratification of rocks? Was such layering of rocks due to a mere playfulness of nature, or was it designed by a Creator, and if so, for which purpose? Two explanatory theories competed around the middle of the nineteenth Century.[51] The German geologist Abraham Gottlob Werner (1749–1817)[52] explained the stratification of rocks by their deposition at the bottom of ancient seas. Earthly matter, kept in suspension in primordial oceans, would slowly settle as a consequence of chemical precipitation,

[48]Hankins, T.L. 1985. Science and the Enlightenment. Cambridge University Press, Cambridge, p. 43.

[49]Chambers, R., 1969 [1844], Vestiges of the Natural History of Creation, de Beer (Ed.). University Press, Leicester; and Humanities Press, New York, p. 40.

[50]Chambers, 1969, ibid., p. 24.

[51]For greater detail, see Gillispie, C.C. 1951. Genesis and Geology. A Study in the Relations of Scientific Thought, Natural Theology, and Social Opinion in Great Britain, 1790–1850. Harvard University Press, Cambridge, MA. Bowler, P.J., 1976. Fossils and Progress. Science History Publ., New York.

[52]Millhauser, 1959, ibid., p. 42.; see also Ospovat, A. 1976. Werner, Abraham Gottlob, pp. 256–264. In: Gillespie, C.G. (Ed.), Dictionary of Scientific Biography, vol. 14. Charles Scribner's Sons, New York.

or of sedimentation under the Law of Gravitation, and would thus continuously contribute to the formation of growing landmasses. His theory addressed the formation of the earth's crust only, not that of the globe in its entirety. Werner's theory was in broad agreement with Leibniz' postulate of a "gradually shrinking ocean,"[53] and seemed to account for the superposition of rocks he had recognized in his home country, Saxony.[54] The oldest rocks, bare of fossils, form the highest mountains; fossiliferous rocks seemed to form mountains and hills of intermediate height, whereas the most recent sediments formed the loose soil of the plains.[55] However, and most importantly perhaps, Werner's theory became entangled in a dispute over the nature of basalt. When Werner first published his *"Short Classification and Description of the Various Kinds of Rocks"* in 1786, the consensus leaned toward an igneous (volcanic) origin of basalt, but Werner maintained his position that basalt was of aqueous origin. When a prize was issued by a natural history magazine for the best essay explaining the nature of basalt, two students of Werner entered the competition, arguing opposing viewpoints. The thesis of the volcanic origin of basalt won the rhetoric of the day, but that did not settle the ongoing controversy. Although never accepted by Werner during his lifetime, the theory of an igneous origin of basalt was eventually shown to be the correct one.[56]

"Neptunism," as Werner's theory was to be named, stood in contrast to "volcanism," a theory put forward by James Hutton (1726–1797),[57] an earlier compatriot of Chambers. Did not every miner experience rising temperatures with increasing depth of the gallery? Georges Buffon[58] had already postulated a source of central heat for planets such as the earth[59] and had even performed experiments with heated copper balls to determine the rate of cooling of the globe to obtain a clue as to what the absolute age of the earth might be. Hutton believed this central heat to be the cause of earthquakes. He accepted the idea that the stratification of rocks was due to the deposition of suspended matter at the bottom of ancient seas, but he did not believe that all continental matter had at any one time been suspended in a primordial global ocean. Instead, he pointed to the erosion of landmasses through the continuous action of water, wind, and gravity, wearing away ancient continents, and transporting earthly matter to the sea for renewed deposition. The "central heat" would bake the sediments into hard rocks, which would be elevated to form new continents through the impact of earthquakes.[60] Hutton viewed the globe as being

---

[53]Rudwick, 1972, ibid., p. 126.

[54]Werner, A.G. 1787. Kurze Klassifikation und Beschreibung der verschiedenen Gebirgsarten. Waltherische Buchhandlung, Dresden.

[55]Rudwick, 1972, ibid., p. 126. For more detail see Ospovat, 1976, ibid., p. 260.

[56]Ospovat, 1976, ibid., pp. 261–262.

[57]Millhauser, 1959, ibid., p. 43.

[58]Millhauser, 1959, ibid., pp. 41–422, 64–65.

[59]Bowler, P.J. 1984. Evolution, the History of an Idea. The University of California Press, Berkeley, p. 32.

[60]For more detail see Eyles, V.A. 1972. Hutton, James, pp. 577–589. In: Gillespie, C.G. (Ed.), Dictionary of Scientific Biography, vol. 6. Charles Scribner's Sons, New York.

subject to continuous destruction and restoration, a historical yet strictly cyclical process that would not support the concept of a directional earth history with a beginning and an end in time.[61] Hutton's earth history, explained in his *"Theory of the Earth"* (1788)[62] moved in circles governed by uniformly acting universal laws of nature; his was a dynamic world, true enough, but again one of dynamic permanence. "We find no vestige of a beginning, no prospect of an end," is a famous line that Hutton issued in paper presented at a meeting of the Royal Society of Edinburgh in 1788. Such a "steady-state" world that would not countenance a directional earth history seemed to Hutton "both scientifically and theologically superior"[63]: it would bring geology closer in line with astronomy, both lines of inquiry aiming at the discovery of universal lawfulness in nature that would reflect back on the Creator.

Charles Darwin had, on the other hand, actually witnessed the elevation of land by an earthquake. Traveling around the world on the HMS Beagle as a gentleman companion of Captain FitzRoy and appointed explorer-naturalist, Darwin's interest were spurred by his reading of Charles Lyell's *"Principles of Geology"* published in three volumes from 1830 to 1833.[64] Lyell, the leading British geologist of his time and a later supporter of Darwin, adopted much of Hutton's "steady-state" world.[65] He again emphasized the uniformity of natural laws: same cause, same effect, always, everywhere, all the time. Causes effective in the past were the same as those we experience today, and will remain the same in the future. An expert in the geology of European mountain chains, Lyell of course acknowledged the facts of erosion and sedimentation, as well as the landscape building powers of volcanic activity. In his view, these forces balanced each other on a global scale. Amidst constant change, everything remained the same. Chambers would read progressive development into the Fossil Record, a view that was later staunchly opposed by Lyell in meetings of the Geological Society of London.[66] In the meantime, however, Darwin entered in his diary on February 20, 1835: "This day has been memorable in the annals of Valdivia, for the most severe earthquake experienced by the oldest inhabitants."[67] A few pages down he went on to point out: "The most remarkable effect of this earthquake was the permanent elevation of land; it would

---

[61]Bowler, 1984, ibid., p. 42.

[62]Hutton, J. 1788. Theory of the Earth; or an Investigation of the Laws Observable in the Composition, Dissolution and Restoration of Land Upon the Globe. Transactions of the Royal Society of Edinburgh, 1: 209–304.

[63]Rudwick, 1972, ibid., p. 179. For a more detailed account see Rudwick, M.J.S. 2008. Worlds Before Adam. The Reconstruction of Geohistory in the Age of Reform. The University of Chicago Press, Chicago.

[64]Lyell, Ch. 1830–1833. Principles of Geology, 3 vols. John Murray, London.

[65]Rudwick, 1972, ibid., p. 179.

[66]Secord, 2000, ibid., p. 418.

[67]Darwin, C., 1962. The Voyage of the Beagle. Annotated and with an Introduction by Leonard Engel. Anchor Books, New York, p. 303.

probably be far more correct to speak of its cause."[68] Exploring the high Andes, Darwin found fossil shells of marine organisms at an altitude of about 1,300 feet, closely resembling shells that could be collected along a sandy beach. The conclusion had to be that earthquakes were responsible for the lifting up of sedimentary rock above sea level. Chambers had read[69] Darwin's account of his voyage on the Beagle, which was first published in 1839 and experienced an immediate success, going through two additional printings within the first year after its publication. Not surprisingly, Chambers found Hutton's "volcanism" better supported that Werner's "neptunism." However, he interpreted the succession of strata from a slightly different angle than Hutton. Chambers had not failed to notice that, according to the geological literature of his time, no remains of organic beings had as yet been found in the lowermost, that is, oldest deposits. The further one moves up the succession of layers of rock, that is, the further one progresses in time from the past toward the present, the more "complex" would the organisms be whose hard parts were preserved as fossils. Such "progressionist geology" outlined by Adam Sedwick, a leading geologist of the time, painted the Fossil Record as an ascent from simple to complex forms of life.[70] To Chambers, this meant that earth history must have had a beginning in time, and from there had progressed in a lawfully predetermined direction toward greater complexity of organization, a process that resulted in the unfolding of the Great Chain of Being through time. But predetermined by whom – and how?

3.6 The First Cause and Secondary Causes

Why should there be no evidence of life on the earth during the earliest phases of its existence? Chambers explained the lack of fossils in early deposits as a consequence of the awesome vehemence of earthquakes supposedly manifest at that time – comparable to the labor-pains of Creation. He was careful, however, not to stipulate any change of the universal laws of nature at any time throughout earth history. Enacted by the Creator, as Chambers believed they were, natural laws were to be eternal and universal, acting uniformly throughout time and space: same cause, same effect, at all times, and anywhere in the universe. The Creator, in Chambers' view, was the First Cause. Residing outside space, time, and matter, the Creator would not move the world along by entering into space-time. Instead, the Creator would enact natural laws, that is, *secondary causes*, to keep the natural processes on earth going, and it is the existence of such laws that for Chambers explained the success of natural sciences in explaining the past and predicting the future. Early earthquakes had more powerful effects not because of a change of the

---

[68]Darwin, 1962, ibid., p. 312.

[69]Hodge, 1972, ibid., p. 130.

[70]Secord, 1994, ibid., p. xxxi.

laws of nature but because the crust of the earth was thinner during those early
phases of the cooling of the globe, and hence less resistant to forceful impacts – a
theory that Chambers borrowed from George Buffon's[71] account of earth history. In
his earth history, first published in 1749, Buffon had invoked so called "revolutions" to account for changes in geology and biology. Yet, he was quick to point out
that such revolutions in earth history would not indicate any changes of the
universal laws of nature. But because the crust of the earth was so much more
delicate during these early times, the same causes, which today take centuries to
have any marked effect, earlier would have resulted in major effects in the course of
years only.

However, once life became manifest on earth, the direction of its evolution was
placated by Chambers by the use of a single, if not simple concept: progress along
the Great Chain of Being. The most primitive conditions of form would characterize early phases of the evolution of life, so called "zoophytes" and corals, living
beings, which even Aristotle had not been able to attribute unequivocally to the
animal or plant kingdom. Trembley's polyp would fall into that category. These
creatures would be followed by brachiopods, worms, and fishes. The fishes
provided the starting point for the conquest of land, first by the imperfect reptiles,
which were, however, superseded by the more perfect mammals. The latest and
most perfect newcomers were humans. Chambers found in the succession of fossils
through the stratigraphic column of successive layers of sedimentary rocks an
expression of the Great Chain of Being[72], a graded succession of increasing
complexity, a mirror image of progressive development pure and simple. Why
did Chambers decide that fishes should supersede annelid worms on the ladder
of life? "The occurrence of annelids is important... They are red-blooded and
hermaphrodite, and form a link of connection between the annulosa (white-blooded
worms) and a humble class of vertebrata", ... "such as amphioxus and myxine" as
Chambers added in a footnote.[73] It was a relation of similarity, judged to be
essential, namely the red color of blood, which motivated Chambers to hypothesize
genealogical relationship.

The classification of organisms, based on the concept of a Great Chain of Being,
mirrored a succession from "lower" to "higher" conditions of life. Living things
succeeded minerals; animals succeeded plants; animals with a backbone (vertebrates) succeeded invertebrates such as worms, mollusks, and echinoderms; tetrapods (animals with four limbs) succeeded fishes; mammals succeeded reptiles. The
Great Chain of Being mirrored a continuous ascent to "improved organization."
Chambers found the same "natural order" to be manifest during the embryonic
development of each organism: "It is only in recent times that physiologists have
observed that each animal passes, in the course of its germinal history, through a
series of changes resembling the *permanent forms* of the various orders of animals

---

[71]Millhauser, 1959, ibid., pp. 41–42, 64–65.

[72]Lovejoy, A.O. 1936. The Great Chain of Being. Harvard University Press, Cambridge, MA.

[73]Chambers, 1969, ibid., p. 62.

inferior to it in the scale."[74] The maggot, he found, resembles the earthworm, the embryonic crab a millipede, the tadpole of a frog would resemble a fish, and the brain of a human embryo would correspond to that of an adult fish. Again, Chambers proved no originality in his argumentation, as he drew on the literature of his time. The parallelism of embryonic development and the Great Chain of Being reaches back to the philosophy of Aristotle, reached the peak of its popularity during the eighteenth century, and was treated as a law of nature by early nineteenth century anatomists such as the German anatomists Johann Friedrich Meckel and Friedrich Tiedemann, and the French author Etienne Serres.[75] The latter in particular had studied the development of the central nervous system in the light of this doctrine, which led Chambers to invoke a "parity of law."[76] By this he meant a correspondence of a universal law of order, manifest in parallel in different branches of natural history such as comparative anatomy, paleontology, and embryology. Classification in terms of the Great Chain of Being, the succession of fossils through the stratigraphic column, as well as the embryonic development of every organism would correspond to the same principle, which is that of progression toward higher levels of organization. As proposed by Chambers, however, this "parity of law" would only address the correspondence of order, but not the causes of such correspondence. What, in fact, were the logical consequences of the recognition of this three-fold-parallelism: the parallelism between the Great Chain of Being, the Fossil Record, and embryonic development?

One basic conclusion became immediately obvious. If it is accepted that the earth, and life on it, had an unidirectional history, and if it is also true that life was impossible on the surface of the globe during the early phases of its evolution, then the conclusion must be that continents and oceans, plants, animals, and humans could not coexist from the very beginning of earth history. Viewed from such a historical perspective, there was no way to escape the conclusion that any creative act of whatever nature – natural or supernatural – has to be extended through time and space, and that it must occur coextensively with the historical development of the earth and its inhabitants. If under these premises Creation was still understood as a Special Act of the Creator, that is, as a direct involvement of the First Cause with natural events and processes, the First Cause would lose its special status as an "unmoved mover" that transcends space and time, but would become subject to the constraints of time and space instead. The result would be an anthropocentric understanding of the First Cause, a Creator sketched in the image of man. Chambers imagined a Creator who would "at one time produce zoophytes, another time to add a few marine mollusks... again to produce crustaceous fishes, again perfect fishes, and so on to the end? This would surely be to take a very mean view of the Creative Power – to, in short, anthropomorphize it, or reduce it to some such character as that

---

[74]Chambers, 1969, ibid., p. 198.

[75]Millhauser, 1959, ibid., p. 73. See also Secord, 1994, ibid, p. xvii.

[76]Chambers, 1969, ibid., p. 71.

borne by the ordinary proceedings of mankind."[77] The eternal Creator could not himself be part of the Great Chain of Being. To pull the Creator down to earth into space and time would mean to picture the Creator in the image of humans.

The First Cause resides outside matter, space and time, yet natural processes are manifest in matter and stretch through space and time: how is it possible to relate one to the other? Chambers' answer was that the Creator would have acted like a monarch, enacting laws that are as time-independent as his own reign. The most basic law amongst those secondary causes that govern the world of living things was, according to Chambers, the "Law of Development." This law governed the propagation of species that is the "reproduction of the like," as Chambers put it. The propagation of species presupposes the reproduction of individuals, and Chambers believed the embryonic development of individuals to repeat the pattern of the Great Chain of Being. Should it happen, as Etienne Serres had envisaged earlier, that an organism fell short of the goal of its individual development that is typical for its species, it would fall back to a lower level of organismic complexity, that is, to a lower rung on the ladder of life. Atavistic malformations were thought to provide the empirical evidence for such a conclusion. However, if a relapse was possible, then its opposite would have to be possible, too. The idea then is that an organism has the potential to develop beyond its species-specific form of life to attain a higher level of complexity, and thus to climb to a higher rung on the ladder of life. Chambers argued that "It is no great boldness to surmise that a super-adequacy... would suffice in a goose to give its progeny the body of a rat, and produce the ornithorhynchus [platypus], or might give the ornithorhynchus the mouth and feet of a true rodent, and thus complete at two stages the passage from the aves to the mammalia,"[78] from birds to mammals. Accordingly, organisms change due to a changing combination of parts. Beyond that, and at Chambers' hands, evolution turns out to be driven by a Law of Development, "to which that of like-production is subordinate."[79] This is the essence of Chambers' "Theory of Evolution," for which he had a famous model, the eighteenth century "science-fiction writer" Benoît de Maillet.[80] In his "*Telliamed*" of 1748[81] (his own name spelled in reverse), this earlier author believed that all land-dwelling animals had their precursors in the sea, and he cited flying fishes and sea elephants as widely known intermediary stages in the "evolution" of birds or elephants. Humans would have originated from "*hommes marins*" (marine humans) as exemplified by the mermaid that had allegedly washed ashore near Amsterdam in 1430.[82] In support of

---

[77]Chambers, 1969, ibid., p. 153.

[78]Chambers, 1969, ibid., p. 219.

[79]Chambers, 1969, ibid., p. 222.

[80]De Beer, 1969, ibid., p. 12. Millhauser, 1959, ibid., pp. 62–63.

[81]Maillet, B. de 1748. Telliamed, ou Entretiens d'un Philosophe Indien avec un Missionaire François. L'Honoré & Fils, Amsterdam.

[82]Maillet, M. de. 1749. Telliamed, ou Entretiens d'un Philosophe Indien avec un Missionair Francois. Librairies Associées, Basle, p. 330, 332.

this thesis he also referred to the notarized testimony delivered by a captain in 1671, who claimed to have spotted a creature half-fish, half-man in the Nordic seas.[83] Small wonder that Darwin, as a reminder not to expose himself to easy criticism or even ridicule, annotated his copy of the *"Vestiges"* (the sixth edition[84]) with the warning: "Never use word higher and lower"[85] when speaking of organisms and their classification.

However, progress is only one side of the coin called evolution – the other side of the coin is adaptation. Nobody could claim that adaptation is progressive. Parasites underwent retrogressive evolution in adaptation to their host environment; snakes lost their limbs in adaptation to their environment. Progressive evolution along the Great Chain of Being and adaptation are two quite independent aspects of species transformation, as was clearly recognized by the early French evolutionist, Jean-Baptiste Lamarck (1744 – 1829), whose *"Philosophie Zoologique"* was published in 1809. Chambers[86] commented on this book, which was generally perceived as the first presentation of a full-fledged evolutionary theory (condemned by its adversaries as an excess of late eighteenth century French materialism).[87] Lamarck was forced to invoke different causal mechanisms to explain adaptation vs. progressive evolution. Adaptation, he believed, was promoted by the use or lack of use of organs: an organ no longer used would atrophy, and would eventually be lost; an organ in constant use would adapt to the functional demands imposed on it, and become modified accordingly. The problem of Lamarck's theory was in trying (in vain) to explain what many of his predecessors and contemporaries believed, namely how such environmentally induced changes of structure, changes induced by the use or lack of use of organs, could become heritable. Imagine a hard-working farmer, who through his labor in the fields develops callosities on his hands. He sometimes wished he had a son who could help him and keep him company through the day. Eventually, the farmer marries and his wife soon after carries his child. It is a son, and given the father's expectations, it would be advantageous for him to be born with callous hands. But his hands are as smooth and delicate as are those of any other babies. There exists no mechanism that renders acquired characteristics heritable.

Progressive evolution, pushing ahead along the Great Chain of Being, was explained by Lamarck on the basis of the action of "subtle fluids." These "subtle fluids" he believed to be inherent in matter, indeed itself of a material nature, albeit a highly elusive one. As the female seminal fluid, this Lamarckian agent of

---

[83]Maillet, 1749, ibid., p. 334.

[84]Secord, 2000, ibid. p. 433.

[85]Bowler, 1990, ibid. p. 107. See also Egerton, F.N. 1970. Refutation and conjecture. Darwin's response to Sedgwick's attack on Chambers. Studies in the History and Philosophy of Science, 1: 180.

[86]Millhauser, 1959, ibid., p. 66.

[87]On Lamarck see also: Burkhardt, R.W. 1977. The Spirit of System. Lamarck and Evolutionary Biology. Harvard University Press, Cambridge, MA. Corsi, P. 1988. The Age of Lamarck. University of California Press, Berkeley.

progressive change again represents a theoretical, that is, unobservable substance that was invoked in an attempt to explain what was believed to be manifest in observation: an increasing complexity of life due to progressive evolution. This atmospheric, "aetheral" substance, once internalized in living matter, would by its continuous and spontaneous circulation through the body drive the latter's development to ever higher levels of increasing complexity and at the same time stimulate the organs to respond to their use or lack thereof. Like Buffon with his "formative force," Lamarck felt entitled to invoke such "subtle fluids" in his explanation of progressive evolution, even though there was no physical evidence for their existence, because in so doing he did nothing else but follow the example set by Newton. Newton famously discovered the fundamental law of gravity, a law that holds across the universe and explains its order and function, but he had no clue as to the material basis of this gravitational force. Consider the "billiard-ball" physics of Newton's time. If any object in motion would hit another one that was at rest, the impact would transmit forces that would cause the relative motion of the two objects to change. This relation, the action of one body on another and the other body's reaction, is perfectly easy to understand, in principle at least. But gravity was a different thing altogether: it postulated forces that act over a distance between two objects. How could this be possible? "Something" had to relate these two objects with each other, transmitting the attractive force. Newton introduced "aether," an elusive substance, to fulfill that role. Lamarck, in turn, postulated lawfulness in progressive development, and called for "subtle fluids" as its material basis. With such illustrious predecessors, Chambers felt fully justified to invoke a universal Law of Development that drives organisms to ever-higher levels of complexity along the Great Chain of Being without a clear understanding of the mechanics of its working.

3.7 A World of Pre-Established Harmony

Like his predecessors, and Darwin after him, Chambers recognized that adaptation could not provide a mechanism to explain progressive evolution along the Great Chain of Being. If progressive evolution did in fact take place, and the Fossil Record as well as embryonic development appeared to him to document this fact, it had to be the consequence of universal lawfulness that would ultimately lead back to the First Cause. On many occasions, however, Chambers marveled at the perfect adaptation of organisms to their particular environment. Because of the action of the Law of Gravity, and subordinate laws of lesser generality, the physical environment was subject to continuous change. And because of the action of the Law of Development, the organic world was subject to continuous change, driving organisms to ever higher levels of complexity. Yet caught in between these continuously changing worlds, one inorganic, the other organic, species were able to maintain perfect adaptation to the physical environment in which they lived. Asking the question of how such harmonious relations could be maintained between organic beings and their physical environment throughout geological time, Chambers concluded that there had to exist a "pre-established" harmony between these two interrelated yet separate realms, both progressing along their own yet parallel trajectories. "Yet, be it remembered, the whole phenomena are, in another point of view, wonders of the highest kind, for in each of them we have to trace the effect of God's Will which had arranged the whole in such harmony with external physical circumstances, that both were developed in parallel steps. . ."[88] According to Chambers' view, the physical world, and that of organic beings, constitute independent systems, yet both of these systems develop in parallel steps, which explains the perfect adaptations of all organisms to their particular environment at all times and through all changes. As the physical environment changes over geological time, so do the plants and animals that populate the surface of the earth. As they change, organic beings would not only maintain perfect adaptation to their conditions of life, but also and simultaneously attain higher levels of complexity.

With his theory of embryogenesis, Buffon had introduced the distinction of organic and inorganic matter. Chambers correspondingly recognized two different and, on his account, independent realms of nature: the inorganic physical environment on the surface of the earth and its living inhabitants. Transformations within the first realm were governed by the universal Law of Gravity, and those taking place within the second realm were governed by the universal Law of Development. Both these laws were secondary causes, enacted by the First Cause in such a way that changes in both realms were, nevertheless, fully in step with one another, although independent from one another. An appropriate metaphor for this vision would be one of two independent clockworks set up in such a way that both would work in perfect synchrony, in perfect pre-established harmony. The notion of a "pre-established" harmony between two independent systems, as introduced by Chambers, requires some more detailed discussion, indeed an excursion into the history of philosophy. Such will provide insights into a number of issues central to Chambers' understanding of humanity's place in nature, which after all is – and always has been – a major issue for every evolutionary theory.

A good way to begin this discussion is with a brief cross-reference to William Paley, whose famous book "*Natural Theology, or Evidences of the Existence and Attributes of the Deity Collected from the Appearance of Nature*" was published in London in the year 1802.[89] Published at the beginning of the nineteenth century, this treatise defined the agenda for Creationism and Intelligent Design in Britain. In his book, Paley set out to accumulate what he considered to be overwhelming evidence for the existence of the First Cause, amassing examples of the ubiquitous perfection of adaptation of organisms to their physical environment. He compared organisms to complex clockworks, the harmonious function of which could not be

---

[88]Chambers, 1969, ibid., p. 223.

[89]Paley, W. 1802. Natural Theology: or Evidences of the Existence and Attributes of the Deity, Collected from the Appearances of Nature. J. Faulder, London.

explained without reference to a plan and purpose underlying their design, thus providing evidence for an intelligent mind underlying Nature. Authors, such as Maupertuis, who argued that accidental variations could add up to the transformation of species, simply had to be wrong by Paley's lights, given what he considered to be overwhelming evidence for design and purpose in nature. Even the slightest change in the organic machinery of a perfectly adapted organism will cause a functional disturbance, as it would in any complicated and integrated clockwork. If organisms were to maintain their perfect adaptation, only two options were left after discounting the accumulation of accidental variation through mechanisms of heredity: one would be the wholesale transformation of species, for which no plausible natural mechanism had (and has) ever been offered; the other was the belief in the immutability of species.

By the time of Paley's writing, the metaphor of clockwork had already had a long history both in philosophy and natural history, fueled by the fondness of the seventeenth century aristocracy in automated playthings of all kind. One seventeenth century philosopher, who not only exerted a major influence on the future development of his field but also went so far as to compare the living organism to an automated clockwork, was the French philosopher, René DesCartes.[90] He not only laid theoretical foundations for physics, but from there proceeded to reduce the natural processes observed in living beings to phenomena explicable by the laws of physics pure and simple. The basic tenet of Cartesian physics was that any body persists in the same mode of motion, and will continue to do so, unless it collides with another body. It is easily appreciated that DesCartes' physics foreshadowed Newton's laws of motion. If some extended body hits another piece of matter, the movement would be transferred from one body to the other, all else being equal. In other words, a body at rest will start to move only if hit by another moving object. This axiom was based on the view that a material body could be defined by its extension in space and time, that is, by the space–time region it occupies, such that no two distinct extended bodies could ever occupy the same region of space at the same time. The movement of an extended body into an already occupied space–time region requires the occupying body to be pushed out of its space–time region by the moving body. The resulting principle of action and corresponding reaction can, supposedly, quite easily explain processes observed to occur between extended bodies, even if these are endowed with life. But here is the problem: "life" at that time was commonly understood as a function not merely of extended bodies such as billiard balls but also of an immaterial soul. For naturalists who opposed the rise of materialism in the mid-eighteenth century, as well as its beginnings at earlier times, it is its vegetative soul that renders a mushroom different from a rock, it is its animal soul that renders the polyp *Hydra viridis* different from a water-plant, and it is the rational soul that renders human beings different from animals. However, if the soul is immaterial, it does not have an extension in space and time, it does not occupy a certain space-time region like a rock that can be kicked away, and hence its function

---

[90]Millhauser, 1959, ibid., pp. 41–42.

cannot be subject to the law of action and reaction. Critics, therefore, upheld that the Cartesian principles of physics would not explain the phenomena of life. DesCartes drew a different conclusion: he perceived a radical duality between soul and body, postulating that the two represented fundamentally different principles or "substances," and that his principles of physics, while explaining the functions of the body, could not explain the functions of the soul. Much later, Gilbert Ryle[91] aptly branded Cartesian metaphysics as the "ghost in the machine" doctrine, the "ghost" being the soul and the "machine" being the body.

To consider the soul and the body as two distinct and different substances, one material, the other "aetheral," is bound to create problems in the explanation of how they interact. The theory of a pre-established harmony between body and soul was first proposed by the German philosopher Gottfried Willhelm Leibniz[92] in his attempt to solve the problem of the duality of "mind and matter." Following the physical principles expounded by DesCartes, he accepted the extension in space and time as essential attributes of matter. While it seemed comparatively easy to account for the transmittance of forces from one body to another, it was less easy to explain the transmittance of any impulse from a material body to the immaterial soul or vice versa. How come that the soul would experience pain, if somebody strikes the body – how does the soul transmit to muscles the impulse to react and strike back? Leibniz accepted the premise that the immortal and hence the immaterial soul would never be able to directly interact with the material body, and on that basis concluded that a complete harmony between the two worlds, the material body and the immaterial soul, had to have been preordained, that is, pre-established by the Creator at the beginning of time. Whenever, in the course of time, any interaction seemed to take place between any body and any mind anywhere at any time, this could not be a true interaction based on physical relations, but had to be a pre-established and as such independent, yet harmonious action of the two substances. The event of somebody striking the body, and the soul stimulating the body to react, would not be explicable in terms of physical interaction between body and soul, but would require, for its proper explanation, a pre-established harmony between the working of body and soul. The Creative Mind, whose handwriting Leibniz believed to be all too evident in the passing of nature, would have reviewed all possible events that could occur in all worlds possible, and would have enacted the best of all possible worlds, predetermined in all, even the most minor and insignificant aspects. The Plan of Creation was a blueprint of the two most complicated clockworks imaginable, one representing the world of matter, the other representing the spiritual realm of the soul, but both running parallel in absolute yet pre-established harmony. The synchronous and harmonious motion of these two clockworks was guaranteed by the universal lawfulness, the secondary causes, that govern the function of the two clockworks. Leibniz proposed his seemingly

---

[91]Ryle, G. 2000 (1949). The Concept of Mind. With an Introduction by Daniel C. Dennett. Chicago University Press, Chicago.
[92]Millhauser, 1959, ibid., pp. 41–42.

extravagant theory of pre-established harmony in opposition to other philosophers of his time, who invoked a spontaneous and special intervention of the First Cause each time any interaction would take place between any mind and any body anywhere in the world. Extravagant at first sight, the Leibnizian system of a pre-established harmony would appear to provide a much more economical explanation for the relation between body and soul than such alternatives of continuous Divine intervention. But beyond the mere issue of simplicity, there was also the need to keep the eternal Creator outside the realm of time and space. And finally, Leibniz furthermore kept the goal of natural science in sight, which is the predictability of the natural course of events. Leibniz' world was one dominated by natural laws, perceived by him to be the only world view supporting the practicability of natural science. Scientific theories that allow successful predictions of future natural processes offered a means to predict, control, even manipulate the natural environment through science. A world that is completely governed by universal laws of nature, that is, by secondary causes, seemed to hold the best prospect for the development of successful natural sciences. In contrast, a world view that allows the First Cause to intervene at any time, anywhere, and in any way with the natural course of events would jeopardize the promise that scientific theories could form the basis for dependable (reliable) predictions.

Similar to Leibniz's proposed solution of the mind–body problem, Chambers claimed that the perfect adaptation of all organisms ever to evolve to their ever changing physical environment had to have been pre-established by the First Cause and was brought about by universal laws of nature that work in a predetermined harmony. The laws enacted by the Creator would ensure that adaptation as well as progressive ascent along the Great Chain of Being would occur in perfect harmony with the ever-continuing change of the physical world. Progression and adaptation were the two aspects of organic evolution, which were predetermined to proceed in perfect harmony with the evolution of the physical environment. For Chambers, there was purpose in nature, which is the perfect adaptation of living creatures to their physical environment. But for him, nature also had a goal, which is the progressive ascent of living beings to ever-higher levels of complexity of organization. In Chambers' "natural system," both purpose and goal conspire to one end: the ascent of humans. Looking at nature from this angle, it became obvious – as previous generations had contended for centuries – that the story of Creation had been written in two languages: in a book of nature and in a book of words. In Chambers' words: "Does it not... appear that our ideas of the Deity can only be worthy of him in the ratio in which we advance in a knowledge of his words and ways; and that the acquisition of this knowledge is consequently an available means of our growing in a genuine reverence for him!"[93]

Chamber invoked two paths to learn about the Creator: through His "words," and through His "ways," that is, through the Holy Bible, and through the study of His

---

[93]Chambers, 1969, ibid., 233.

Creation. The "metaphor of the two books"[94] is a strong one, motivating the development of natural sciences that would eventually try to describe natural phenomena in the time-less languages of logic and mathematics. To the degree that this is possible, the workings of the human mind would appear to match the workings of nature, governed by laws of nature, that is, secondary causes. Newton, again, is the glowing example, whose mind had disclosed the inner workings of the universe. For Chambers, this was just another expression of the Law of Development, which would not only ensure the development of increasing intellectual capabilities over time, but also the perfect adaptation of those intellectual capabilities to the physical environment. But such a view breaks down the barrier that had been claimed to separate humans from animals. Motivated by his Law of Development, Chambers tore down the radical distinction between an animal soul, proper to animals, and a rational soul, an essential property of humans: "There is, in reality, nothing to prevent our regarding man as especially endowed with an immortal spirit, at the same time that his ordinary mental manifestations are looked upon as simple phenomena resulting from organization, those of the lower animals being absolutely the same in character, though developed within much narrower limits."[95] Humans would become part of nature: "Man, then considered zoologically, and without regard to the distinct character assigned to him by theology, simply takes its place as. . . true and unmistakable head of animated nature upon this earth."[96] This view of life, however, had serious consequences. Chambers did not believe that humans had been "specially endowed with an immortal spirit."[97] Hence there was no *essential* difference between humans and animals, a difference that would radically distinguish man from animal in all and every one of Leibniz' possible worlds. In a gradually developing, continuously evolving world, there could be no gaps, there could be differences of degree only. For Chambers, "The difference between mind in lower animals and in man is a difference in degree only; it is not a specific difference!"[98]

Chambers' "*Vestiges*" has been characterized as a book that grounded cosmology as much as natural history in a natural and progressive developmental process.[99] Chambers aspired to deliver a vision of the animated world that would be on par with the grandiose vision of the physical world as had been developed by Newton, yet presented in a style accessible to a broad audience. "*Vestiges*" sketches a materialistic view of this animated world: there remain mysteries to be explained, such as the material basis for the actions of the Laws of Development. But if such explanations remained elusive, the mysteries nevertheless were locked up in matter

---

[94]Schweber, S.S. 1989. John Herschel and Charles Darwin: a study in parallel lives. Journal of the History of Biology 22: 3.

[95]Chambers, 1969, ibid., p. 326.

[96]Chambers, 1969, ibid., pp. 272–273.

[97]Chambers, 1969, ibid., p. 326.

[98]Chambers, 1969, ibid., p. 335–336.

[99]Secord, 1994, ibid., p. ix.

and with it in time and space. The residence of the eternal Creator remained beyond time and space, just as the secondary causes enacted by Him transcend time and space. Universal Laws of Nature would insure the harmonious progressive development of the inanimate and animated world, maintaining a pre-established harmony between the two realms. The fact that science works, that natural processes unfolding in time and space can be captured in terms of universal natural laws, expressible in the time-less language of mathematics as Newton had shown, was in Chambers' view nothing but another expression of the pre-established harmony between the realms of physics and biology. The human being, endowed with the power of rational reasoning, is not thereby set apart from nature, but instead is embedded in nature as its crowing production.

Chapter 4
Stemming the Tide of Change

Opposing all theories of species transformation, including that of Chambers, Hugh Miller appealed to Aristotelian metaphysics: nature is pervaded by design and purpose. But design requires a designer, purpose requires a planning mind. Miller also appealed to the empiricist tradition in the philosophy of science: the proper foundation for science is unbiased observation, not idle speculation. Chambers had appealed to the continuity of nature: the unbroken Great Chain of Being was paralleled in the Fossil Record and in embryonic development. In contrast, Miller maintained that even if there were an unbroken succession of fossils through the Geological Record, there was still no process of species transformation to be observed, but rather only a certain 'top upon bottom order of things'. And it was precisely this pattern which revealed design, purpose and goal-directedness in the Fossil Record through the occurrence of fossils of a 'prophetic type'. The characteristics of such 'prophetic' fossils announced at an earlier epoch of earth history the later appearance of new forms of life.

Design, purpose, and goal-directedness were also apparent, for Miller, in embryonic development, which preserved the perfect adaptation of species to the place in the household of nature to which they had been assigned by the Creator. Species cannot undergo gradual transformation, as this would disrupt their perfect adaptation to their environment. There consequently must be gaps in nature separating not only species from one another, but also man from beast. The rejection of the concept of 'perfect adaptation' was likely the most difficult intellectual hurdle that Darwin faced in developing his theory of evolution. Perfection cannot vary, but wherever Darwin investigated nature with enough concern for detail, he found organisms that form a population to be subject to variation. Organisms were not perfectly adapted, but adequately adapted – adequate for survival. This insight provided Darwin with the key to his theory of natural selection.

4.1 Hugh Miller (1802–1856)

On October 10, 1802, Hugh Miller was born in Cromarty, Scotland. His father, a proud owner of a fishing schooner, drowned when Hugh was only 5-years old. [1] At school, the boy drew attention neither by his diligence nor by his accomplishments.

---

[1]Rudwick, M.J.S. 1974. Miller, Hugh, p. 388–390. In: Gillispie, C.C. (Ed.), Dictionary of Scientific Biography, Vol. IIX. Charles Scribner's Sons, New York.

O. Rieppel, *Evolutionary Theory and the Creation Controversy*,
DOI 10.1007/978-3-642-14896-5_4, © Springer-Verlag Berlin Heidelberg 2011

He preferred to roam in the Scottish landscape, strolling through the fields and hills and along the seaside, thereby experiencing the first self-taught lessons in natural history, which proved so productive in the later years – as his spiritual father, the paleontologist Louis Agassiz, was to profess.[2] It was not easy for the adolescent to submit to the reality of life; however, in February 1821, he finally decided to become a stonemason, working in the quarries of northern Scotland. The quarry where he found employment was located close to Stromness on the Orkney Islands. The workers there excavated the Old Red Sandstone, deposited at the bottom of an ancient Sea during the Devonian, approximately from 416 to 359 million years ago.[3] Unaware of the actual time involved, the accredited scientists at the time, nevertheless, believed that these deposits of early age were devoid of any signals of ancient vertebrate life. "Geologists of high character had believed that the Old Red Standstone was defective in organic remains; and it was not till after 10 years, acquaintance with it that Mr. Miller discovered it to be *richly fossiliferous.*"[4] Miller unearthed the remains of bizarre creatures, fishes of an as yet unknown structure. These discoveries not only gained Miller the appreciation and support of one of the leading paleontologists and zoologists of his time, Professor Louis Agassiz, but also stimulated him to communicate his insights to a wider public. His book *"The Old Red Sandstone"* was published in 1841, a text that by no means was restricted to the enumeration of properties characteristic of these strange early fishes, but which really was intended as a polemic against early theories of evolution, in particular against the theory sketched in 1809 by Jean-Baptiste Chevalier de Lamarck[5], renowned invertebrate zoologist and paleontologist at the National Museum of Natural History in Paris.

Given Agassiz' role as spiritual mentor of Hugh Miller, it is of importance to shed some more light on this person and his background.[6] Although his work on glaciers that resulted in the concept of an Ice Age was important, as can be judged from Chambers' public support for it (see the previous chapter), it was not the main research focus of Agassiz. His real love was for fishes, both living and fossil. First, he got involved with research on fishes when he was invited to work on collections from Brazil that had been brought back by Johann Baptist von Spix, who had deposited them in the Natural History Collections in Munich. Having completed this task in 1829, Agassiz embarked on a description of the fish fauna in lake Neuchâtel (Switzerland), quickly expanding his scientific interests to the fossil fishes from the black shales of Glarus (Switzerland), and from the limestones of

---

[2]Agassiz, L. 1850. Hugh Miller, Author of 'Old Red Sandstone' and 'Footprints of the Creator, pp. xi-xxxvii. In: Miller, H. 1849 [1850], Foot-Prints of the Creator: or, the *Asterolepis* of Stromness, Agassiz, L. (Ed.). Gould, Kendall and Lincoln, Boston.

[3]The Geological Society of America 2009 Geological Time Scale, http://www.geosociety.org/science/timescale/

[4]Agassiz, 1850, ibid., p. xxi.

[5]Bowler, P.J. 1976. Fossils and Progress. Science History Publications, New York, p. 53.

[6]Winsor, M.P. 1991. Reading the Shape of Nature. Comparative Zoology at the Agassiz Museum. The University of Chicago Press, Chicago.

Monte Bolca (Italy). During these early phases of his career, young Louis Agassiz remained undeterred in his attempts to make the acquaintance of the famous Georges Cuvier from the National Museum of Natural History in Paris. At the zenith of his career, during which he became one of the most influential zoologists, paleontologists, and geologists of the early nineteenth century, Cuvier was not an easily approachable man. But by the end of 1831, Agassiz has established himself in Paris as a student of the great man.[7]

4.2 Georges Cuvier and Louis Agassiz: Experts on Fossil Fishes

Georges Léopold Chrétien Frédéric Dagobert Cuvier[8] had joined the Paris Museum after the French Revolution, when, under Napoleon Bonaparte, he embarked on a stunning career, which took him not only to the position of the leading comparative and functional anatomist of his time, but also to the privileges and power of France's Secretary of Education. Cuvier is said to have worked at seven desks simultaneously, his investigations spanning the entire series of vertebrate animals from fossil fishes to fossil mammals. Behind his back he came to be nicknamed "the mammoth," not only because of his increasing *gravitas* that translated into a massive bodily *Gestalt* but also because of his many scientific achievements, which included the conclusive proof for the extinction of fossil species such as the mastodon and the mammoth.[9] Cuvier was a typical representative of the conservative restauration that followed the revolution in Paris. He rejected all ideas of species transformation, as he remained the last surviving proponent of the doctrine of the encapsulation of preformed and preexisting embryos, all created at the beginning of time.[10] In a celebrated presentation to the National Institute of France in 1796, Cuvier argued that Buffon was dead wrong when he claimed that the present day elephant represents a "degenerated," that is, transformed mammoth. Instead, Cuvier was the first to show that today there exist two species of elephants, the African and the Asian one, and that the mammoth is a third, separate species that

---

[7]Lurie, E. 1960. Louis Agassiz: A Life in Science. The University of Chicago Press, Chicago.

[8]Rieppel, O. 2001. Georges Cuvier (1769–1832), pp. 139–156. In: Jahn, J., and M. Schmitt (Eds.), Darwin & Co., Vol. 1. C.H. Beck, Munich, and references therein.

[9]Bourdier, F. 1971. Cuvier, Georges, p. 524. In: Gillispie, C.C. (Ed.), Dictionary of Scientific Biography, Vol. 3. Charles Scribner's Sons, New York. See also Abel, T. 1987. The Cuvier-Geoffroy Debate. French Biology in the Decades before Darwin. Oxford University Press, Oxford. On Cuvier in general see also: Daudin, H. 1926. Cuvier et Lamarck. Les Classes Zoologiques et l'Idée de la Série Animale. Félix Alcan, Paris. Rudwick, M.J.S. 1972. The Meaning of Fossils. Macdonald, London. Rudwick, M.J.S. 2005. Bursting the Limits of Time. The Reconstruction of Geohistory in the Age of Revolution. The University of Chicago Press, Chicago.

[10]Coleman W. 1964. Georges Cuvier, Zoologist. Cambridge MA: Harvard University Press, pp. 162ff.

went terminally extinct instead of having been transformed into an elephant. Such an extinct species Cuvier called *"une éspèce perdue,"*[11] a "lost species," that is, one that had irretrievably disappeared from the surface of the earth. Cuvier thus became the first to prove the fact of extinction. If a fossil mollusk shell did not match with that of any living mollusk, it was at that time still possible to claim that the species documented by fossils was still extant. But because its occurrence might be restricted to some small bay along the coast of some remote island, it might have escaped the collecting efforts of past and present explorers. In contrast, nobody could claim that the mammoth known from fossils still existed somewhere, for surely such a large mammal could not have escaped its discovery.

Through his functional anatomical studies, Cuvier hardened the doctrine of the functional correlation of parts, which rendered an atomistic conception of organisms as an aggregate of originally separate parts impossible, and which further emphasized the perfect adaptation of species to their specific environment. To him, such a perfectly adapted organism was a complex machine that could not be changed part by part without its function being compromised. Such a complex clockwork would either have to remain the same, or to be redesigned from ground up and transformed all at once, which evidently was impossible. For Cuvier, the Law of the Functional Correlation of Parts was the most important law that governs the animated world, a law of such high rank and universal generality that even the almighty Creator would submit Himself to its reign. The reason is that Cuvier thought that this law would impart on the living world a necessity on par with that expressed by mathematical laws[12]: "In one word, the form of a tooth indicates the form of the jaw joint, the latter implies the structure of the palate, which in turn presupposes the structure of the claw, in the same way as the mathematical equation of a curve implies all is properties."[13] It was on the basis of this law that Cuvier believed that he could reconstruct the whole organic machinery of an animal known only from an incompletely preserved fossil. "The most insignificant facet on a bone, the most weakly expressed apophysis, determine the class, the order, the genus and the species to which this bone belongs. This is true to the degree that if we only have the well-preserved extremity of a limb bone available... we can identify it with as much certainty as if we had the whole animal at our disposal."[14] There was no room for gradual, piece-meal change of organic forms in Cuvier's world, no room for accidental "mistakes" in the coming together of parts that would form an embryo. Species, such as the mammoth, may go extinct, but they do not change. It was Lamarck, overseeing the collections of invertebrate animals other than insects, who along with Etienne Geoffry Saint-Hilaire supported Cuvier's appointment at

---

[11]Cuvier, G. 1828. Discours sur les Révolutions de la Surface du Globe, 5th Ed. G. Dufour and Ed. D'Ocagne, Paris, p. 117.

[12]Coleman W., 1964, ibid., p. 67.

[13]Cuvier, 1828, ibid., p. 98f.

[14]Cuvier, 1828, ibid., p. 105.

the Paris Museum.[15] Having demonstrated the extinction of fossil mammal species, Cuvier expected his senior colleague, Lamarck, to develop equally conclusive proof for the extinction of fossil mollusk and cephalopod species. But Lamarck drew quite different conclusions from his studies of fossil invertebrates. According to him, the successive appearance of different species in the Fossil Record would not document the extinction of species and their miraculous replacement with new ones, but instead the transformation of older species into different, new ones. This was enough to turn Cuvier from a supporter into an enemy. Cuvier did not even refrain from expressing his complete disagreement by ridiculing Lamarck's ideas in the eulogy he held at the latter's funeral in 1829.[16]

In contrast, Cuvier was so impressed by Agassiz' work on fossil fishes that he left his own notes on the subject to the young adept.[17] After Cuvier's death in 1832, Agassiz returned to Neuchâtel, where he became a professor in natural history at the *Lyceum*. Throughout his life, Agassiz held up and defended Cuvier's ideas and theories about the earth and its inhabitants. And just as Cuvier remained, till the end of his life, a staunch opponent to early theories of transformationism such as Lamarck's, so did Louis Agassiz oppose Darwin's theory of evolution till the end of his life. In one of his reviews of Darwin's work dating from 1874, Agassiz concluded that "however broken the geological record may be, there is a complete sequence [of fossil species] in many parts of it", and yet – in spite of this – "there is no evidence of a direct descent of later from earlier species in the geological succession of animals."[18] This echoes the theme of Hugh Miller's book *The Old Red Sandstone* from 1841, which was designed to highlight the putative flaws that are incurred when species transformation is inferred from the Fossil Record. Miller was concerned that authors might draw illegitimate conclusions from the sequence of fossils that marks out earth history. But why should it be illegitimate to conclude from a graded series of superimposed fossils to a theory of species transformation? Agassiz' and Miller's arguments here reflect a strict adherence to a now defunct empiricist philosophy that was dominant at their time, one that admitted sensory experience alone and nothing else as the sole basis of secure, scientific knowledge. True enough, a graded series of superimposed fossil forms may be suggestive of species transformation. But since species transformation itself was not a process observable in the Fossil Record, corresponding theories could in Agassiz' and Miller's views not claim a sound scientific basis. Indeed, and before too long, Miller should find his concerns in that regard fully justified: Chambers' *"Vestiges of Natural Creation"* was published in 1844. Miller, who had by that time become the

---

[15]Appel, 1987, ibid, p. 12.

[16]Appel, 1987, ibid., pp. 168f.

[17]Andrews, S.M. 1982. The Discovery of Fossil Fishes in Scotland up to 1845. Royal Scottish Museum, Edinburgh, p. 7.

[18]Hull, D.L. 1973. Darwin and His Critics. The Reception of Darwin's Theory of Evolution by the Scientific Community. Harvard University Press, p. 445.

frontman of the Creationist camp[19], set out to deal transformationism another and hopefully definitive blow with the description of yet another fossil from the Old Red Sandstone, a fish he identified as belonging to the genus *Asterolepis*. Because of the author's health problems, the book entitled *"Foot-Prints of the Creator or, the Asterolepis from Stromness,"* appeared somewhat delayed in 1849. The title Chambers had chosen, *"Vestiges,"* derived from the Latin term *vestigium*, meaning a "trace" or a "mark", such as a "foot-mark."[20] Miller set out to retrace the footprints of the Creator in a different light than Chambers had shone on them.

Miller's interpretation of fossils becomes a lot more intelligible when viewed in the light of an early but crucial experience he had during his career as a stonemason, an experience that was recounted by Agassiz in his introduction to a later edition of Miller's book on *Asterolepis*.[21] The first blasting Miller witnessed in the quarry exposed two dead birds, hidden in a deep fissure of the quarry walls. Obviously, the birds had sought shelter during a recent storm that had devastated the Orkney Islands off the coast of northern Scotland. This incidence led the self-taught paleontologist to the erroneous conclusion that fossils, remains of once living organisms found in the succession of layers of rocks that form the earth's crust, need not to have lived and strived where they are found today. And indeed, there are circumstances where fossils are found in so-called fissure fillings. A fissure may have developed in ancient bedrock, to be filled during a later geological epoch with sediment carrying fossils. The age of the fossils will then not correspond to that of the bedrock, but to that of the fissure-filling sediment. But Miller's conclusions were more radical: the incidence indicated to him that the mere stratigraphical succession, i.e. the successional occurrence of fossils through geological time, was not in and of itself enough to determine ancestor–descendant relationships. While the latter claim is correct in the light of modern paleontology, the former claim is not. Birds caught in a fissure between rocks is not the same thing as fossils embedded in sedimentary deposits, irrespective of whether these are filling ancient fissures, or – as is normally the case – superimposed one on top of the other. As early as 1666, the Danish scientists, Niels Stensen, working in Florence and generally known as "Steno," had raised and answered the question how it could be possible for one solid body, such as a fossil shark tooth, to be fully encased in another solid body, such as a chunk of rock? The answer: the shark must have lived, died, and decomposed (or shed its tooth as a function of tooth replacement) before the sedimentary rock that encloses it had formed. The tooth must have sunk into the sediment, and become fully enclosed in it before the sediment compacted to form solid stone.[22]

---

[19]Secord, J.A. 2000. Victorian Sensation. The Extraordinary Publication, Reception, and Secret of *Vestiges of the Natural History of Creation*. The University of Chicago Press, Chicago, p. 280.

[20]Secord, 2000, ibid., p. 104.

[21]Agassiz, 1850, ibid.

[22]Poulsen, J.E., and E. Snorrason. 1986. Nicolaus Steno, 1638–1686.Nordisk Insulin Laboratorium, Gentofte, Denmark. See also Rudwick, M.J.S. 1972. The Meaning of Fossils. Macdonald, London, pp. 49ff.

4.3   Miller's Attack on Transformationism

Miller opened his defense of creationism by attacking the transformationist idea endorsed by Chambers that humans could have emerged from the animal kingdom by gradual progressive evolution through a series of intermediate stages. Such an outlook would imply that the distinction of humans from animals was no longer a principal, an *essential* one, but only a matter of degree. There would be no essential property left that distinguished man from beast. This, according to Miller and many of his contemporaries, would undermine the whole codex of morale and ethics founded on a belief in Christian values, which in his view provided the foundation of human society. In making this point, Miller sided with eminent authorities of his time, such as the Reverend Adam Sedgwick, Woodwardian Professor of Geology at Cambridge University, who in his review forcefully rejected Chambers' *"Vestiges"* on the grounds that the new cosmology negated all distinctions between the realm of physical objects and that of moral values.[23] Miller could not agree more: "It is the fact that man must believingly coöperate with God in the work of preparation for the final dynasty, or exist throughout its never-ending cycles as a lost and degraded creature, that alone renders the development hypothesis formidable."[24] The destruction of the argument of human descent from the animal kingdom necessarily had to start with a clear refutation of the thesis of the continuity of progressive development, believed by transformationists to bridge the gap between animals and humans. Here again, Sedgwick showed the way. The problem is, once again, one of change.

Ancient Greek atomists thought that the changing objects of nature were formed by the coming together of eternal, unchanging, and indestructible atoms of different kinds in different combinations at different times. But for these atoms to come together, and separate again, required them to move through space. And for that to be possible, the atomists claimed the existence of "empty space" through which the atoms could move. In contrast, Aristotle thought that nature abhors empty space. According to him, the vacuum does not exist. From the rejection of "empty space" followed the famous "Principle of Continuity": everything is connected in an unbroken series or chain of forms or events. There are no gaps in nature, neither in terms of empty space, nor in terms of interrupted processes. Darwin himself would reach the same conclusion in his *"Origin"* of 1859: *"Natura non facit saltum"* – nature does not make jumps.[25] According to Darwin, both the animated and inanimate realms of nature are subject to gradual, instead of saltational change. Just as earthquakes build mountains in a long series of small steps, and erosion

---

[23]Secord, J.A. 1994. Introduction. In: Secord, J.A. (Ed.), Robert Chambers: Vestiges of the Natural History of Creation, and Other Evolutionary Writings. The University of Chicago Press, Chicago, p. xxxii.

[24]Miller, 1850. ibid., p. 336. See also Bowler, P.J. 1984. Evolution. The History of an Idea. The University of California Press, Berkeley, p. 139; Rudwick, 1972, ibid., p. 207.

[25]Darwin, C. 1859. On the Origin of Species. John Murray, London, p. 194.

would gradually flatten out mountains over time, species would gradually transform in small steps over long periods of time. Nature forms an unbroken continuum without gaps or empty space. The natural flow of events was thought to be governed by uniform and universal laws of nature. A gap, or empty space, in nature would disrupt the governance of natural processes by uniform laws of nature. A geometrical line, a spatial distance, the flow of time, a spatially extended body can all logically be subdivided into an infinite number of segments or parts. The same is true for a continuous process of change: the continuity of any process can be subdivided into infinitesimally small segments, the explanation of any one of which through natural causes would stretch nobody's imagination. Darwin's experience, mentioned in the preceding chapter, of the earthquake of Valdivia, provides a classic example: The earthquake had elevated the land, but to such a minor degree that only the local population, acquainted with every detail of the natural environment, would recognize it: "I could discover no evidence of this fact, except in the united testimony of the inhabitants, that one little rocky shoal, now exposed, was formerly covered with water."[26] A minimal uplift of the coastline was the effect, and its cause was a single earthquake of natural dimensions. This observation should not evoke much surprise or skepticism. But it could also be adduced to explain a second observation made by Darwin of fossil seashells "found at a height of 1,300 feet" above current sea level.[27] "A bad earthquake at once destroys our oldest associations: the earth, the very emblem of solidity, has moved beneath our feet like a thin crust over a fluid."[28] A succession of such events stretching through deep time could change the face of the earth dramatically. One simply has to allow nature enough time to complete its works: the successive accumulation of small effects of a natural magnitude, each caused by natural causes such as an earthquake, could dramatically reshape the landscape if allowed to play out over a long enough period of time. Enormous was the power of argumentation based on the principle of continuity, and although not yet pronounced with the sophistication and power of persuasion commanded by Darwin at the time of his writing, Miller correctly identified the force of Chambers' arguments that were built on the continuity of the Great Chain of Being and on the underlying lawfulness linking each level of organization in a gradual and progressive process of change that runs parallel to the lawful development of the physical universe.

Indeed, only a slow, gradual, and stepwise process of transformation would be amenable to an explanation that links natural causes to equally natural effects which each could be viewed as quite plausible in a chain of natural events governed by universal laws of nature. The tale of a human being with six fingers and toes might evoke some surprise and interest, the tale of a human being with six arms and six legs would evoke an incredulous stare. Buffon would not countenance deMaillet's

[26]Darwin, C., 1962. The Voyage of the Beagle. Annotated and with an Introduction by Leonard Engel. Anchor Books, New York, p. 312.

[27]Darwin, 1962, ibid., p. 312.

[28]Darwin, 1962, ibid., p. 303.

idea that humans arouse from mermaids and their male counterparts, but he found it plausible to think that the donkey might have originated from the horse through a process of degeneration. Maupertuis speculated that the accumulation of minor heritable changes could lead to the origin of new species, and so did Darwin, a century later. Any criticism of theories of species transformation had to target the thesis of continuity in nature. The documentation of major discontinuities would leave gaps in the chain of natural events governed by secondary causes, and open the door to the intervention of the First Cause, to the possibility of a spontaneous Act of Creation. Miller was a skilled polemic, and early on in his book he pointed to the fact that the transition from life to death was by no means a gradual one. Death, like birth or pregnancy, does not come in degrees. One cannot be dead and alive at one and the same time, just as one cannot be pregnant and not pregnant at one and the same time. Claims to the contrary would run up against the law of noncontradiction. Consequently, a continuity of such transitions that would create fuzzy boundaries could not be reconciled with logic and rational argumentation, no more than other transitions such as, for example, from a dead rock to a living coral, or from a morally indifferent animal to a rationally thinking and morally sensitive human being. However, rhetoric may be good enough a weapon to try and win a battle, but is generally insufficient to win the war. Miller consequently set out to defeat the transformationists using their own weapons on the battlefield of natural sciences.

If it were true, as Chambers had claimed that the succession of fossils through time would parallel the unfolding of complexity through the process of embryonic development, or the ascent to increasing complexity as reflected by natural classifications that place mushrooms at the bottom, humans at the top, then one would have to predict that early fossils would resemble the embryos of higher organisms, i.e., of organisms of greater complexity that appear later in the Fossil Record. Even Louis Agassiz had claimed that the class of fishes was still in an embryonic stage of differentiation during the early phases of its occurrence in the Fossil Record (i.e., in the Devonian).[29] Quarrying the Old Red Sandstone of Devonian age, Miller had dug up a strange creature that was named *Asterolepis*, at that time one of the earliest known fossil fishes, but it was neither small, nor of simple structure. It was a large animal with a highly complex anatomy, which, in some of its features such as the structure of its teeth, Miller believed to "foreshadow" the age of reptiles, animals that Miller erroneously believed to share a similar tooth structure as his newly found fossil. Using the current time scale and modern knowledge of the Fossil record, *Astereolepis* lived about 390 million years before the present, whereas the earliest uncontested reptiles appeared around 315 million years before the present. It is crucial at this point to recognize the implications of Miller's claim that the tooth structure seen in this archaic fish, *Asterolepis*, would foreshadow the later ascent of reptiles. This could only mean that the much later appearance of reptiles

---

[29]Agassiz, L. 1844. Monographie des Poissons Fossiles du Vieux Grès Rouge ou Devonien (Old Red Sandstone) des Iles Britanniques et de Russie. Jent et Gassmann, Soleure, p. 18.

was already announced by the earliest fossils of vertebrate animals and that the eventual appearance of reptiles on the surface of the earth therefore had already been planned and preordained – something that could certainly not be explained with reference to unconscious and "blind" laws of nature.

Again it was Louis Agassiz who had first introduced the concept of "prophetic types,"[30] which during early phases of earth history would foreshadow what was to follow in what he thought represented the unfolding of the Plan of Creation through geological time.[31] Thus, marine reptiles or ichthyosaurs, widespread during the Mesozoic, were thought to represent a type of organization announcing the later appearance of dolphins, a group of mammals; flying reptiles of the order *Pterosauria* would prophesize the later appearance of birds. No genealogical relationship was implied in the notion of "prophetic types", however, only a correspondence of type of construction and of mode of living, that is, a correspondence of the functional and ecological role these types would play at different geological times in the household of nature. The concept of "prophetic types" could make sense only if one was willing to agree that foresight and goal-directedness underlie the succession of fossils through time, in other words, that there existed a blueprint of Creation. Any claim that the structure of early fossils would hint at later forms of life must presuppose the existence of a rational and conscious agent who had thought through, planned, and executed the Creation of organismic diversity and its successive manifestation in the course of geological time. Only conscious planning could impart a direction and goal, such as progress, on an unfolding process, culminating in the appearance of humans. If *Asterolepis* could be shown, as Miller (erroneously) believed, to share certain similarities in tooth structure with reptiles, it could be regarded as a "prophetic type," and as such would lend support to the belief in a goal-directedness of nature. Yet, on that account, what in the Fossil Record may appear like a progression from one "type" of organization (fishes) to another (reptiles) through time does not correspond to any real process of change. Like the oak tree that preexists in the acorn, the "prophetic type" preexists in an atemporal, ideal world, a world beyond matter, space, and time. The "type" is rooted in Divine thought; Divine thought is revealed by the succession of fossils. The "prophetic type's" appearance on the surface of earth is only its becoming instantiated, that is, its becoming revealed in an instance or an example located in space and time – in this case in the form of *Asterolepis* from Stromness. It announced the arrival of the reptiles in a later geological epoch. But that did not imply that reptiles evolved from fishes through intermediary amphibians. It only meant that the type of construction of a reptile, just only hinted at in the teeth of *Asterolepis*, would be fully revealed through examples at a later epoch of earth history. Ancient reptile species, such as the species of non-avian dinosaurs so

---

[30]Winsor, M.P. 1976. Starfish, Jellyfish, and the Order of Life. Yale University Press, New Haven, p. 101, 148.

[31]Lurie, E. 1960. Louis Agassiz: A Life in Science. The University of Chicago Press, Chicago, p. 162.

popular today, did go extinct. They are irretrievably lost in deep time. The "type" of construction of a reptile cannot go extinct: it existed before the reptiles appeared in the Fossil Record, and it would persist even if all reptiles had vanished from the surface of the earth. The "type" of a reptile, again, does not exist in nature, only its exemplars do. For Agassiz and Miller, the "type" of a reptile was grounded in Divine thought. Although the succession of fossils might suggest to the naïve observer a progression that results from change through time, nothing really changes in Miller's view. The world may appear as a dynamic one to the human observer, but it is really only one of dynamic permanence.

"*Vestiges of the Natural History of Creation*" proclaimed a progressive development of organisms on the basis of natural laws (i.e., secondary causes) rather than through Special Creation (i.e., as a direct effect of a First Cause). The reason for Chambers was that a Creator wrapped up in time and space would cast an anthropocentric view of the Deity. In order to support his claim for a progressive process of change in nature, he had to link the living organisms in an unbroken chain of increasing complexity. One example, mentioned in the previous chapter, was Chambers' use of the red blood color as a character that links annelid worms with vertebrates, relating the two groups in a series of ancestors and descendants. But that was not the most important character he turned to. The main feature on which Chambers based his evolutionary conclusions was the progressive development of the brain through the vertebrate series.[32] In Chambers' view, brain size steadily and continuously increased from fish through reptiles and birds to mammals and humans. In his critique of Chambers, Miller agreed that there was an increase in complexity of the brain as could be shown by calculating and tabulating relative brain size of vertebrate animals. In fishes, the brain was calculated by Miller to be about twice the volume of the spinal cord; in reptiles and birds, the ratio had increased to 2.5:1–3.5:1; mammals showed a ratio of 4:1; however, humans displayed a ratio of 23:1. Viewing these figures, as they obtained on Miller's account, who could argue persuasively for a continuity of transition between animals and humans? The Plan of Creation was progressive, but not continuous: on Miller's account, there was a jump, a discontinuity in brain size between mammals and humans. In the very same sense, there was for Miller a striking discontinuity between even the highest mammals and humans, as it is only the latter that are endowed with the powers of rational thinking, language, and morality. Both the vegetative soul that regulated growth, as well as the animal soul that regulated movement and sensibility, may well be inherent in matter. In contrast, humans become endowed with a rational soul through divine intervention. This is what creates the gap between humans and animals that was appealed to by Adam Sedgwick and, in his tow, by Hugh Miller.

In Miller's view, continuity of progressive transformation was further denied by the fact that the earliest representatives of any particular group need not also be the most simple in structural terms. *Asterolepis* was the perfect example, representing

---

[32]Bowler, 1976, ibid., p. 81.

one of the earliest fossil fish then known, which, nevertheless, was characterized by a relative brain size estimated by Miller – erroneously, but befitting a "prophetic type" – to be comparable to that of reptiles. And finally, Miller marshaled the old argument of retrogressive development, again following in the footsteps of Sedgwick[33]: progress was not the only attribute of organismic diversity reflected by the Fossil Record, embryonic development, and classification – there were also incidences of rudimentation. For example, snakes have too many characters in common with lizards to classify them as anything else but reptiles. Yet, as Agassiz was to show in great detail in his *Essay on Classification*, published a little less than a decade later in 1857, this implied that snakes must have lost limbs, which are generally present in the group of which they form a part (i.e., reptiles). Snakes were to be characterized not by the absence of limbs but by the loss of limbs through retrogressive embryonic development. Borrowing in a roundabout and indirect way from the theories of the great French anatomists Etienne Geoffroy Saint-Hilaire and Etienne Serres, who had developed the "*loi de compensation*," the law of compensation[34], Miller pointed out that snakes compensated for the loss of limbs by an increase in the number of vertebral elements during their embryonic development.[35]

4.4 Miller's Concept of "Proper Science"

Like many of his contemporaries, Miller considered the significance of the Fossil Record an important issue in the creation debate. At the same time, however, there existed for Miller a constant threat that science could be corrupted by the social and political agendas of scientists.[36] To oppose such tendencies, Miller proposed definitions for scientific subdisciplines, which would serve the purpose to concentrate research efforts to the objects and goals of those particular fields of interest. Remember that Chambers' "*Vestiges*" was perceived as an account of natural history that drew from a great variety of special sciences such as astronomy, geology, and biology. Its author was consequently castigated for trying to break down the conceptual and instrumental barriers that separated the various subdisciplines, which professional scientists were engaged in. "*Vestiges*" simply came across as eccentric writing compared to the standard scientific practice of the day.[37] In support of such professionalization through the compartmentalization of science, Miller defined the task of comparative anatomists to determine with accuracy the level of complexity at which organisms had to be classified in the

---

[33]Secord, 1994, ibid., p. xxxi.

[34]Miller, H. 1849 [1850], Foot-Prints of the Creator: or, the *Asterolepis* of Stromness, Agassiz, L. (Ed.). Gould, Kendall and Lincoln, Boston, p. 119.

[35]Miller, 1850. ibid., p. 181–182.

[36]Hull, D.L. 1988. Science as a Process. The University of Chicago Press, Chicago.

[37]Secord, 2000, ibid., p. 201

Great Chain of Being. The task of the paleontologist was to document the temporal succession of organisms in the Fossil Record, and nothing more. Like the science of the anatomist, paleontology was – on Miller's definition – concerned not with causal explanation but with order in nature. The great British empiricist philosophers such as David Hume (1711–1776) and John Stuart Mill (1806–1873) had argued that all sound scientific knowledge ultimately derives from sensory, in particular perceptional (observational) experience. Accordingly, an empirical approach to the Fossil Record consisted in the documentation of the stratigraphic succession of fossil species, based on observation that must be free of all theory and preconception. This is what the paleontologist was called upon to do according to Miller's definition of this field of research. To stretch theory construction beyond such a documentation of the occurrence of fossils in successive layers of rock would be nothing but idle speculation. The result of paleontological research has to be a catalogue of the spatial and temporal distribution of extinct species, and not theories about their possible transformation that remain without observational support. The Fossil Record shows beyond any doubt, however, that not all levels of organization known today existed from the beginning: there were times when fishes cruised the oceans, but no amphibians or reptiles populated the continents. And again: amphibians and reptiles existed long before mammals, and mammals existed long before the appearance of humans. And yet, for Miller, this was not an effect of species transformation at all. Instead, and as indicated by the existence of "prophetic types," what seemed to be a newcomer in earth history is merely the coming into actual existence through exemplification of a "type" of animal construction that has neither past nor any future but just is in the form of a template, of a blueprint of Creation that is not tied to time and space.

Miller maintained that the simple fact of temporal succession of the appearance of fossils was, in itself, far from a valid or even significant proof for genealogical relationships. The nature of fossils had, indeed, remained controversial over a long period of time. As already mentioned, it was Steno who, in a dissertation dating back to 1669, pointed out that sedimentation, the natural settlement of suspended matter at the bottom of water bodies, was the obvious process by which hard tissues of once living animals such as shark teeth could become enclosed in successive layers of sedimentary rocks. This insight paved the way to the understanding that underlying strata of an undisturbed sequence of deposits must be geologically older than overlying strata. Folding and thrusting of the earth's crust might alter the original sequence of strata, a job for geologists to find out, but the basic "Law of Superposition" must still hold, since it is subsumed by the Law of Gravity: the geologically younger deposits, and the fossils they contain, lie on top of underlying deposits in the sequence of the stratigraphical column that date from an earlier time, just as long as the sequence of layers has not been altered by some geological (tectonic) events. But the superposition of strata and the fossils they contain does not also and always imply that earlier fossils are ancestors of later forms of life. This is because the Fossil Record will remain incomplete to a certain degree. Recognizing a fossil as the earliest known representative of its group does not, at the same time, imply that this fossil is also the most primitive member of its group, ancestral to all the later representatives of the same group. An even earlier, and more primitive representative of the same group may have existed, but may simply not have been found (yet). *Asterolepis*, one of the earliest fossil fish known to Miller, shows a highly complex anatomy. But that, in itself, did not necessarily refute Chambers' "Law of Development," for it could always be argued that still earlier, more primitive fishes had existed, but had not yet been found. This, indeed, proved later to be the case.

For Miller, such an appeal to the incompleteness of the Fossil Record was a leap of faith, not knowledge backed up by observation. In the absence of evidence, any claim could be made in hypothetical support of any theory anybody wants to be true. Miller countered forcefully[38] that "The *possible* fossil can have no more standing in this controversy than the "*possible* angel." What he meant by this was that the Fossil Record had to be taken for what it *is*, and not for what it *could be*. Unbiased, simple, and straightforward observation, he contended, would reveal *patterns* of similarity among fossils, but nowhere was a *process* of descent with modification to be seen. Miller introduced a farmer into this hypothetical discourse[39], "a plain, observant, elderly man," who – as a consequence of these virtues – simply cannot follow the expositions his companion, an enlightened philosopher who wants him to accept the theory of species transformation. The philosopher is convinced that the temporal succession of fossils in itself provides sufficient evidence for the development of life on earth: "Look here. . . life, both vegetable and animal, first began," he explained as he pointed to the farmer's own "deep ditch" – and from its beginning species continuously transformed to reach ever higher levels of complexity. The farmer, however, characterized by Miller as being less inclined toward speculation, maintained that he could not *see* anything else but a "certain top-upon-bottom order of things." Unbiased, simple, and straightforward observation is what Miller wanted his science to be based on, and such a science did not, in his view, support the hypothesis of species transformation.

At first sight, this objection seems rather trivial, or else simply wrong. Trivial, because the process of descent with modification is supposed to have taken place in the past, and therefore has to be inferred from the patterns of greater or lesser similarity observed among the fossils distributed over time and space. Wrong, because Miller was appealing to the possibility of theory-free observation, something that modern philosophers of science have found to be impossible. The philosophy of science that started out with David Hume's claim that all knowledge comes from unbiased observation is called empiricism, and Miller insisted that paleontology, the study of fossils, has to be founded on empiricism as has to be the case for any other respectable science. The issue of empiricism will necessitate a more detailed discussion, as does the significance of "observation" in the construction of scientific theories. At this juncture, it may suffice to point out that Miller insisted on "raw" observation that yields brute sense-data, which, to his knowledge,

---

[38]Miller, 1850, ibid., p. 241.

[39]Miller, 1850, ibid., p. 233–235.

and to the date of the publication of his book, had never revealed any *process* of transformation from one fossil (or living) species to another. All that was recognized at that time was a constancy of form of species, and a succession of different species through time. Was there a way to bring the issue of species transformation into closer focus, to seek a way to confirm or disconfirm species transformation through observation? In empirical support of his views, Miller quoted the mummies brought back to Paris by Etienne Geoffroy Saint-Hilaire, who had accompanied Napoleon Bonaparte on his unsuccessful campaign to incorporate Egypt into the French empire.[40] Thousands of years old, these mummies indicated no essential change of human anatomy compared to Miller's compatriots. The absence of anatomical change had likewise been demonstrated by Georges Cuvier in his study of Hibiscus skeletons that had been retrieved from Egyptian tombs[41] and brought back to Paris by Geoffroy. It is ironic that Geoffroy Saint-Hilaire was instrumental in the appointment of Cuvier at the Paris Natural History Museum, when later in their life, they would be divided about issues of comparative anatomy, animal classification, and the potential for species transformation. Cuvier took the evidence at face value: species do not change over time, and if different species are observed at different time horizons, this cannot be due to species transformation. This was also the point of view endorsed by Agassiz and Miller. The reason for Cuvier to take this stance was, of course, his "Law of the Functional Correlation of Parts." With a carnivore dentition goes a carnivore's intestinal system; with an herbivore dentition goes an herbivore's intestinal system. There can be nothing in between. Organisms are complex machines that cannot change in any partial or gradual, stepwise manner. Cuvier had so much confidence in the functional correlation of parts that he proceeded to a public demonstration of the predictive power of what he believed to be a natural law. A fossil from the quarries of Montmartre had only been partially freed from the surrounding sediment, yet enough for Cuvier to recognize it as the skeleton of a marsupial mammal. Cuvier proceeded to enact a public presentation of the further preparation of the fossil that was to expose the pelvic girdle. Once completed, the procedure confirmed Cuvier's previous prediction that the pelvis would include a marsupial bone.[42] Geoffroy Saint-Hilaire, on the other hand, reached the conclusion that species could change, within limits, in adaptation to the physical conditions of their life. However, he reached this conclusion not through the study of the Egyptian mummies he had brought back to Paris, but through the study of fossil crocodiles.[43] This, of course, allowed a

---

[40]Rieppel, O. 2001. Etienne Geoffroy Saint-Hilaire (1772–1844), pp. 157–175. In: Jahn, J., and M. Schmitt (Eds.), Darwin & Co. Vol. 1., C.H. Beck, Munich; and references cited therein.

[41]Miller, 1850, ibid., p. 278.

[42]Rudwick, 1974, ibid., p. 116.

[43]Geoffroy Saint-Hilaire, E. 1825. Recherches sur l'organisation des Gaviales, etc. Mémoires du Muséum d'Histoire Naturelle, 12: 97–155. Geoffroy Saint-Hilaire, E. 1833a. Troisième Mémoire. . . des recherches faîtes dans les carrières du Calcaire Oolithique de Caen, etc. Mémoires de l'Académie Royale des Sciences de l'Institut de France, 12: 44–61. Geoffroy Saint-Hilaire, E. 1833b. Le degree d'influence du monde ambient pour modifier les formes animals; question

much longer stretch of time for transformation to play out. Living crocodiles are characterized by a unique specialization in their skull, which is a secondary palate. What this means is that the internal nostrils (choanae) are displaced backwards to a posterior position in the palate, similar in some loose sense to their (independently evolved) position in mammals. The internal nares thus come to lie behind a skin flap that can close off the buccal cavity from the air passage, such that the animals can breathe with their head partially submerged in water – the typical stalking position of crocodiles. What Geoffroy found in fossil crocodiles was a more anterior position of the internal nostrils, a position that more closely corresponds to the primitive reptile condition. What is furthermore the case is that the developing embryos of living crocodiles show a progressive shift of the internal nares from a primitive anterior position, corresponding to the condition observed in fossil crocodiles, to a posterior position. The embryonic development thus again revealed a parallelism with the Fossil Record. Geoffroy naturally concluded that crocodiles had changed through time as a consequence of species transformation.

4.5   Karl Ernst von Baer and the Importance of Embryology

In contrast, constancy of structure and discontinuity between structural types, this was the message that Cuvier, Agassiz, and Miller took to the public in their argument against transformationism. Chambers had likewise realized that there were gaps to be bridged, or to be explained away, in order to preserve the absolute continuity of the Great Chain of Being. To a reader of his *"Vestiges,"* he may appear to have done so by a sleight of hand. Charles Babbage, an English mathematician he referred to, had been working on the construction of a calculating machine. The project was to build a machine that would be programmable so that it would switch, suddenly and apparently discontinuously, from one calculating operation to another without special external input, but solely on the basis of the program that was entered beforehand. The operations would have to appear to be discontinuous if the machine was observed from the uninformed perspective of an outsider, but in fact they would be in perfect accordance with the natural course of events as determined from preordained laws. In a similar vein, Chambers maintained that the secondary causes that govern the natural course of events would do so in a uniform and continuous manner, even though a human observer might perceive discontinuity. Etienne Geoffroy Saint Hilaire, who had come to accept the idea that species could have the potential for change within the limits set by their general structural type, had provided an interesting example where continuity of cause could result in discontinuity of anatomical structure.[44] Reptiles and birds appear to a certain

intéressant l'origine des espèces téléosauriennes et successivement celle des animauux de l'époque actuelle. Mémoires de l'Académie Royale des Sciences de l'Institut de France, 12: 63–92.
[44]Geoffroy SDaint-Hilaire, 1833b, ibid., p. 80.

degree to represent a similar structural plan, one that was later recognized as the sauropsidan body plan. Yet, in adaptation to flight, the bird lung had achieved a structural complexity that far exceeded the complexity exhibited by any reptile lung. As far as the anatomy of the lung is concerned, there appeared to be a striking discontinuity between a reptile and a bird. However, if one inspected the lung in an embryonic lizard, and in an embryonic bird, a striking similarity of structure would be revealed. There could be no question also that the continuity of the process of embryonic development in both lizard and bird was governed by a continuity of the underlying causal chain. Geoffroy's conclusion was that a very minor change in the early embryonic development of a bird could lead to a vastly different anatomy of the lung in the adult, but such a discontinuity in the adult structure of a lizard and a bird lung did not necessarily imply a discontinuity of the underlying causal chain that governs the embryonic development of lizards and birds.

With his contemplations on how a reptile lung could have transformed into a bird lung, Geoffroy invoked a model of embryonic development that was fundamentally different from the one advocated by Chambers in some parts of his "*Vestiges.*" To see embryonic development as running parallel to the Great Chain of Being is to claim that organisms of higher complexity recapitulate, during their development, the adult anatomy of ancestral forms, which stand on a lower rung of the ladder of life. At some point of its development, the frog tadpole resembles a fish. Geoffroy's claim was quite different. What he said was that embryos of reptiles and birds look closely similar at early stages of their development, but then become increasingly dissimilar as development proceeds. Historians of biology call the first model one of recapitulation, the second model one of differentiation.[45] On the first model, the embryo recapitulates during its development the Great Chain of Being up to the placement of its own species on the ladder of life. On the second model, embryos that resemble each other during early phases of their development become increasingly more different from one another as they grow and differentiate.[46] This second model of diverging development was championed by the eminent developmental biologist Karl Ernst von Baer, who in his monumental monograph of 1828 challenged the idea that embryonic development runs parallel to the Great Chain of Being. In fact, Chambers alluded to both models of embryonic development in his "*Vestiges.*" To go from a goose to a platypus to a rodent is a march along the Great Chain of Being. To find the embryos of fishes, reptiles, birds, and mammals to be in an "identical condition"[47] during early stages of development but to progressively diverge from one another during subsequent stages is a von Baerian model of differentiation, which Chambers picked up from William Carpenter, who was a

---

[45]Nyhart, L.K. 1995. Biology Takes Form. Animal Morphology and the German Universities, 1800–1900. The University of Chicago Press, Chicago.

[46]Richards, R.J. 1992. The Meaning of Evolution. The Morphological Construction and Ideological Reconstruction of Darwin's Theory. The University of Chicago Press, Chicago.

[47]Chambers, R., 1969 [1844], Vestiges of the Natural History of Creation, de Beer (Ed.). University Press, Leicester; and Humanities Press, New York, p. 212.

friend of Darwin[48] and among the earliest authors to defend von Baer's views in England.[49] That the two models of embryonic development, recapitulation vs. von Baerian differentiation, are quite different is best brought out in graphical representation. Illustrating "recapitulation" would be a sketch of a ladder, an unbroken chain of ascending forms of organization. To illustrate "differentiation" one would have to sketch a branching diagram, a tree-like structure with gaps between the tips of its terminal branches, as was indeed done by Chambers.[50] It may seem puzzling that Chambers juxtaposed two distinctly different, even contradictory models of embryonic development in his "*Vestiges,*" as this could be predicted to invite criticism. But James A. Secord[51] thinks that there might have been advantages from a strategic point of view, as von Baer's work remained rather poorly known in Britain.

Von Baer's was an ingenious idea: apparent discontinuity of adult form did not mean discontinuity of developmental process. In fact, how to bring what appears to be discontinuity of form together with continuity of developmental processes is currently a very active field of research in evolutionary developmental biology. But Miller had, of course, another ax to grind. Assume Geoffroy was right, and assume that minor changes in early embryonic development could indeed result in drastic changes of adult anatomy, how then would it be possible that the developmental process would be programmed such that the changed adult form was perfectly adapted to a changed locale in the household of nature? How, in other words, was it possible that the developmental process in birds was programmed such as to result in an adult lung structure that would be perfectly adapted to the demands of active flight? The concept of "perfect adaptation" so invoked obviously implies some foresight, planning, and design, as William Paley had pointed out in his "*Natural Theology*" of 1802. How is this notion of design to be understood? In Paley's case, as well as in Miller's, quite literally: engineers design structures, such as bridges, or clockworks, according to a blueprint, for a certain purpose, and hence toward a certain goal.

4.6   Intelligent Design and the Four Aristotelian Causes

The search for design, purpose, and goal-directedness in nature goes back to Ancient Greek philosophy. Natural processes in general are understood as events, or chains of events; events, in turn, link a cause with an effect. Science seeks to

---

[48]Oppenheimer, J. 1968. Embryological enigma in the Origin of Species, p. 309. In: Glass, B., O. Temkin, and W.L. Strauss jr. (Eds.), Forerunners of Darwin, 1745-1859. Johns Hopkins University Press, Baltimore.

[49]Secord, 1994, ibid., p. xvii.

[50]Chambers, 1969 [1844], ibid., p. 212.

[51]Secord, 2000, ibid., p. 106.

unravel the causes that underlie natural processes, and – classically – to capture those in the form of natural laws. In contrast to modern science, the Ancient Greek philosopher Aristotle had distinguished four causes, which can perhaps best be explained by invoking a carpenter who sets out to build a cabinet. The wood and the nails the carpenter would need are called the *material cause*, as they provide the material with which to build the cabinet. The force with which the carpenter's arm would drive the saw and the hammer is called the *efficient cause*, as it provoked an effect on the wood being cut and nailed together. However, in order to build a cabinet, the carpenter would be well advised to plan his actions ahead of time, for example by drawing up a blueprint of how the finished cabinet should look like. Aristotle called this blueprint, the plan according to which to build the cabinet, the *formal cause*. Finally, the cabinet will only be useful if its construction is planned with reference to the future intended use of this piece of furniture; this intended use of the cabinet, the purpose for which the cabinet is being built, that is, the goal of the whole project Aristotle called its *final cause*. It is easily understood how the concept of perfect adaptation accommodates an Aristotelian conception of causality that implies purpose, goal-directedness, and hence design. Almost all naturalists before Darwin viewed organisms as perfectly adapted to their environment. They seemed to be made to perfectly fit into a particular place in the household of nature, into a particular niche of their environment. And the process of individual development (i.e., the process of ontogeny that comprises both embryonic and postembryonic development) of an organism appeared to replicate and maintain the perfect adaptation of its species through generations: ontogeny looked as if it were goal-directed for a certain purpose, which is the maintenance of perfect adaptation of the developing organism, according to a certain plan, which is the body-plan or *type* of the species to which the developing organism belongs. Such purposefulness and goal-directedness of natural processes was, according to Hugh Miller, imparted on those by the First Cause. According to Chambers, such natural processes were governed by secondary causes, in particular by his Law of Development. There was no need for a supra-natural agent to direct the natural course of events; that direction was, according to Chambers, inherent in nature, naturalized as a consequence of secondary causes enacted by the First Cause. But direction there still was governed by design and directed toward a goal – even for Chambers.

Formal and final causes are inextricably linked to the notion of intelligent design that results in perfect adaptation. Intelligent design in turn is inextricably linked to a rational entity capable of foresight and planning. For Chambers, the world was composed of two perfectly designed clockworks, constructed and wound up by the Creator, or First Cause, but left on their own to unwind according to plan. The Creator does not reside in nature. He cannot reside in nature if nature is subject to change through time, as argued by Chambers. In Chambers' system, the formal and final causes were naturalized in the workings of secondary causes. Miller rejected such a naturalization of creative forces. He found such naturalization unsupported by "raw" observation that is free of idle speculation, and he found it objectionable on moral grounds. Steeped in the evangelical tradition[52], he wanted the Creator more intimately involved with His Creation. His "was the voice of Old Dissent,"[53] which would not allow to insert secondary causes between God and Nature.

Looked at from this perspective, it is easily understood that one, if not *the* major, obstacle that Darwin had to overcome in his thinking was the notion of "perfect adaptation."[54] Darwin struggled to free his mind from the concept of "perfect adaptation" to make room for the variation of organisms that would in turn offer natural selection the opportunity to shape new, and different, forms of life. Darwin noted that animal breeding could result in imperfection, such as in hairless dogs that have imperfect teeth[55], he found hybridization to result in imperfect reproductive organs[56], and perhaps most notoriously, he found "numerous gradations" to link the "perfect and complex eye to one very imperfect."[57] A developmental stage of cirripedes, which are barnacle crustaceans, Darwin described as having "six pairs of beautifully constructed natatory legs, a pair of magnificent compound eyes, and extremely complex antennae; but they have a closed and imperfect mouth, and cannot feed."[58] Taking the concepts of design and perfect adaptation to communities of co-existing species results in further complications. Consider the coexistence of carnivores, such as lions, and herbivores, such as antelopes. Lions would presumably be perfectly adapted to hunt and capture antelopes; antelopes would presumably be perfectly adapted to escape lions. Something in this system has to give to keep either species from going extinct, lions as a consequence of starvation (because the perfect antelopes would always escape them), antelopes as a consequence of predation (because the perfect lions would always catch them). So how is it that lions and antelopes can coexist, as a result of a certain balance between predator and prey? There seem to be only two options: a perfectly balanced, preordained and preestablished harmony between predator and prey, enacted by the First Cause, or competition. While our old friend from previous chapters Charles Bonnet – among others – opted for the first solution[59], Darwin opted for the latter explanation.

Most important for Darwin, though, was what he came to call the "Great Law of Variation." Again, Darwin had no viable theory of inheritance available as an explanation for variation when he published his *Origin* in 1859. His "Law" therefore is what philosophers of science call an "empirical generalization," a generalization based on repeated observation. And indeed, wherever Darwin looked with

[52]Secord, 2000, ibid., pp. 279, 282.

[53]Secord, 2000, ibid., p. 206.

[54]Ospovat, D. 1981. The Development of Darwin's Theory. Natural History, Natural Theology & Natural Selection 1838–1859. Cambridge University Press, Cambridge.

[55]Darwin, 1859, ibid., p. 12.

[56]Darwin, 1859, ibid., p. 262, 264.

[57]Darwin, 1859, ibid., p. 186.

[58]Darwin, 1859, ibid., p. 441.

[59]Bonnet, Ch. 1764. Contemplation de la Nature. Marc-Michel Rey, Amsterdam.

enough concern for detail, he found that in all plant and animal populations no one individual organism looks exactly the same as any other one. This was true of the beetles he collected while still a student, of the birds he collected on the Galapagos Islands during his voyage on the Beagle, of the barnacles he studied back at home, and of the orchids he liked to grow. Variation spoils the concept of perfect adaptation: perfection cannot vary. There can only be one way to be perfect, and this perfection would be embodied in the specific "type" that is exemplified by each species – in its blueprint. Species, it turns out however, are variable, and hence cannot be perfectly adapted; instead, they are adequately adapted for survival, adequate enough for continued participation in the processes of reproduction, variation, and natural selection. If there were a blueprint specifying the "type" of a species, this blueprint would have to be blurred.

Through his many contacts with plant and animal breeders, Darwin assured himself that the variations he recorded in nature were heritable. Indeed, by selecting from the variants for further breeding, plant and animal breeders mimicked under artificial conditions the process of selection that Darwin eventually proposed also to operate in nature under natural conditions. In 1868, Darwin published his *"Variation of Plants and Animals under Domestication,"* a book in which he finally came up with his theory of inheritance he called *"Pangenesis."* Darwin invoked heritable particles that he called "gemmules" or *"gemmulae,"* which in his theory performed in much the same way as Buffon had claimed for his "organic molecules." Circulating through the parental body, the gemmulae would be imprinted by this organism's characteristics, including characters that the parental organism had acquired during its lifetime (here, Darwin fell victim to the same error that had previously marred Maupertius' understanding of inheritance, as well as Lamarck's theory of evolution). Eventually, the imprinted gemmules would be carried through the blood stream to the reproductive organ for storage. Upon conception, both parental organisms, male and female, would contribute gemmules to the formation of the embryo. This is, again, an atomistic conception of embryogenesis, where parts come together to form a new organismic whole. Since the embryo was formed from minuscule particles, and since these particles could be variously recombined in the formation of offspring, the newly formed organism could vary in ever so slight degrees in every one of its parts and particles. Such *fundamental* variation of organisms that collectively form an interbreeding population is the raw material on which, according to Darwin, natural selection operates. Those variants that have even only a slight edge over others in their adaptation to their natural environment would see this to manifest itself through a relatively greater reproductive success, even if only slightly so. This is not a "nature red in tooth and claw," as Alfred Tennyson poetically put it. It is variable organisms competing for relative reproductive success, which would see slightly better adapted traits to be more frequently reproduced than less well-adapted traits. Species transform in the course of an extended process of variation and natural selection, interconnected by innumerable transitional forms that together form a continuous, unbroken chain of generations.

This was a theory of species transformation, indeed a theory of the origin of new species, which was far more radical than Chambers'. It was a theory that severed the ties to Aristotelian metaphysics, one that rejected the call for *formal* and *final* causes. As we shall see, it was an altogether different approach to science in general, and to biology in particular, than that marshaled by either Chambers or Miller. Indeed, at the end of the day, Darwin would not only have offered a viable theory of evolution but would also have redefined what science is, not necessarily for physics and astronomy but certainly for historical biology. It was a science free of design and purpose, a science where certainty was replaced by probability. But even if the theory of variation and the consequent theory of natural selection were one of numbers and of statistics, respectively, for Darwin these theories were still Laws of Nature.

How did Darwin arrive at this solution? Clearly, the debate between Chambers and Miller – one that Darwin closely studied – had brought a great number of issues to the forefront, some of greater, some of lesser generality. The most important perhaps is the question: what is "respectable science"? How is science to be organized, departmentalized, or synthesized? What role does observation play in science, and how far can theory construction be allowed to transcend observation? What is, quite generally speaking, the proper mode of scientific reasoning? What are secondary causes (i.e., universal laws of nature), how are they discovered, and what are they supposed to explain: unobservable causal relations or merely observed regularity? All of these questions, relevant to science in general, have implications for the more specialized issues that Chambers and Miller were debating. Is there order in nature that can be discovered, or is order in nature merely the reflection of an ordering human mind? And if there is, indeed, order in nature, what does it mean: is it anchored in a blueprint of creation, or is it evidence for evolutionary relationships? What does it mean to have an orderly succession of fossils through the layers of rock that encode earth history? What does it mean that embryos of broadly related groups of organisms, such as the vertebrates, are similar during early stages of development, but become progressively more dissimilar as they grow and differentiate? These are the questions we now must turn to.

Chapter 5
A Matter of (Natural) Laws

Science seeks to discover laws of nature. Historically, such laws of nature were understood as 'secondary causes', enacted by a 'First Cause'. Such an understanding of natural laws resulted in the 'metaphor of the two books'. The Creator revealed himself through the Book of Revelation as much as through the Book of Nature. Exhaustive knowledge of the laws of nature would lead all the way back to knowledge of the First Cause. On this account, the study of nature would provide insights into moral and ethical issues. The attempt to derive what 'ought to be' from what 'just is' constituted a major roadblock preventing the acceptance of a materialistic conception of nature in general, and of theories of species transformation based on nothing but natural causes in particular.

Laws of nature, enacted by a First Cause, would be universal laws, imparting necessity on the natural course of events. Universal laws are timeless, and so would be nature that is governed by them. The paradigmatic science investigating such a world of dynamic permanence was astronomy. To subject biology to the doctrine of dynamic permanence required a doctrine of pre-existence: the paradigmatic example was the butterfly, pre-existing in the caterpillar. But how did the world, and its inhabitants, come into being in the first place? The 'Kant-Laplacian nebular hypothesis' called for initial chaos, out of which the order of the universe emerged under the guidance of natural laws. Here, the First Cause, and with it ethical norms and moral principles, slipped out of science and the world it explained. Nature was left to organize itself, knowledge of the laws of nature rendered science reliably predictive. But in order to be reliably predictive, natural laws must link causes to effects. In contrast to the laws of physics, Chambers' Law of Development was not predictive, but only descriptive. It described a three-fold parallelism of order in nature, apparent in the Great Chain of Being, in the Fossil Record, and in embryonic development, but it could not afford predictions based on causal relations.

Darwin had to overcome a number of intellectual hurdles before he arrived at his 'Law of Natural Selection'. He had to abandon the metaphor of the two books that placed moral values and ethics into nature; he had to abandon the concept of the Great Chain of being and the associated three-fold parallelism; and he had to abandon the requirement of universality for natural laws. The astronomer John William Herschel characterized the theory of natural selection as the 'law of the higgledy-piggledy', because it made explicit reference to time as it obtains from the passing of nature. But by adapting the nature of scientific explanations to the requirements of historical biology, Darwin was able to account for genuine change through time: the origin of new species from ancestral ones.

O. Rieppel, *Evolutionary Theory and the Creation Controversy*,
DOI 10.1007/978-3-642-14896-5_5, © Springer-Verlag Berlin Heidelberg 2011

5.1   The Metaphor of the Two Books

The book that made Paul Feyerabend famous was called "*Against Method*" (1975).[1] Paul Feyerabend (of whom more later), a philosopher of science, who called himself a "methodological anarchist," did not believe in the possibility of defining science. It is, therefore, no surprise that he took issue with such definitions, especially if these were proffered in a controversial socio-political context, such as the trial of Creation Science in court. How about defining science as "that which scientists do, or what is accepted by the scientific community"? This can be read as a very weak definition of science, one that was interpreted by Feyerabend as not necessarily excluding "Creation Science" from genuine sciences. Although, as we shall see later, this definition has quite a lot going for it, it does seem at first sight to leave us with a rather deflated view of science. A more specific definition of science might seek to ground scientific knowledge in the causal relations that govern the world of experience. If scientific theories deal primarily with natural causes and their effects, then science appears to be motivated by the desire for the discovery of laws of nature. This, then, is the second definition of science cited by Feyerabend[2], one that again was given in the context of the lawsuit against the Arkansas Act 590 of 1981, which required balanced teaching of evolutionary theory and "creation science" in Arkansas public schools.

Accordingly, science

1. Is driven by Natural Laws
2. Must be explicable with reference to Natural Laws
3. Is testable with respect to the experienced (empirical) world
4. Draws conclusions that are defeasible (may turn out to be wrong) and are not *ultimate*
5. Is falsifiable

In short, science seeks to formulate the relations between natural causes and their effects in terms of natural laws. It is the desire for the discovery of natural laws that motivates scientists to work late hours. The laws that scientists write about must be able to allow testable predictions, and must be at least potentially falsifiable. But what, exactly, is such a "Natural Law"?

Chambers found the world to be governed by two fundamental laws of nature: the inanimate world was governed by the Law of Gravity and the animate world by the Law of Development. Whereas it was the First Cause that Chambers believed to have enact these laws, only the laws of nature, the secondary causes, were amenable to scientific investigation. His view of science did not motivate Chambers believed

---

[1]Feyerabend, P. 1975. Against Method. Outline of an Anarchist Theory of Knowledge. Verso Edition 1978, London.

[2]Feyerabend, P. 1982. Auszug aus dem Urteil des Distriktrichters gegen das Land Arkansas vom 5. Januar 1982, pp. 227–230. In: Feyerabend, P., and C. Thomas (Eds.), Wissenschaft und Tradition. Verlag der Fachvereine, Zürich.

to abandon his belief in a First Cause; it was just that the First Cause was not, indeed could not be, subject to scientific investigation. Chambers was not alone in adopting the "metaphor of the two books" that allowed two avenues for the study of the First Cause: one is the Book of Revelation (the study of the Holy Bible), the other the Book of Nature (the study of lawfulness in nature). During the first half of the nineteenth century, the metaphor of the two books was enshrined in the educational curriculum at the universities of Cambridge and Oxford[3], although neither Chambers nor Miller graduated from either of these two famous schools. Hugh Miller's adoption of the metaphor is clearly documented by his distinction between the knowledge of a geologist ("We know, as geologists. . .") on the one hand, which is the knowledge of a natural scientist investigating the temporal succession of fossils (the "*Fossil Record*"), and the Book of Revelation on the other hand, which is the "*Revealed Record.*"[4] However, there is an important consequence of this metaphor of the two books: it is that the secondary causes, which underlie the laws of nature, render these laws universal. These are laws, which are true everywhere and at all times, laws that are exceptionless and thus impart *necessity* on the natural course of events. For the philosopher Gottfried Wilhelm Leibniz, the metaphor of the two books delivered an important promissory note: if the secondary causes are universal laws of nature, then the exhaustive knowledge of the laws of nature would ultimately lead all the way back to knowledge of the First Cause, which enacted those secondary causes in the first place.

The "metaphor of the two books," as any other, has its history. The medieval philosopher Thomas Aquinas (1225–1274) tried to convey to his contemporaries the importance of the ethical standards and moral values that were to be derived from the scientific investigation of nature.[5] He believed that the study of what "is" in nature, enacted as it were by the Creator, teaches the student of natural history what "ought to be": moral values and ethical standards could be derived from the study of nature. Johannes Kepler (1571–1630), the leading astronomer of his time, declared the pursuit of natural science a "worship in the temple of nature"[6]; he took the harmony of celestial movements as evidence of the universality and rationality of the First Cause.[7] In a letter, he characterized astronomers as priests of the supreme Deity, celebrating the Book of Nature.[8] Galileo Galilei (1564–1642) understood nature in comparable terms as a transcript of the Book of Revelation, written in the timeless language of mathematics. The idea that the study of natural laws would reveal the handwriting of the Creator and that the study of nature would

---

[3]Schweber, S.S., 1989. John Herschel and Charles Darwin: a study in parallel lives. Journal of the History of Biology 22:3.

[4]Miller, H. 1849 [1850], Foot-Prints of the Creator: or, the *Asterolepis* of Stromness, Agassiz, L. (Ed.). Gould, Kendall and Lincoln, Boston, p. 325.

[5]Rolfes, E. (Ed.) 1977. Die Philosophie des Theomas von Aquin. Felix Meiner Verlag, Hamburg, pp. 41, 43.

[6]Hemleben, J. 1971. Kepler. Rowohlt, Reinbeck bei Hamburg, p. 93.

[7]Hemleben, 1971, ibid., p. 93.

[8]Hemleben, 1971, ibid., p. 53.

thus reveal to the investigating mind normative moral and ethical values that are to guide human behavior, constituted one of the most fundamental roadblocks for the acceptance of materialist theories of change in nature. If it is true that blind forces inherent in matter drive species transformation, if it is true that humans originated through such a process of species transformation, then humans would seem to carry the ugly baggage of their ancestry: a beastly nature conditioned by blind chance. Recall the famous Reverend Adam Sedgwick, a critic of Chambers' vision even harsher than Miller, who chastised *"Vestiges"* for having *"annulled all distinction between physical and moral."*[9] This is why Miller, and many of his contemporaries, rejected Chambers' Law of Development that bridges the division between human and beast. To abandon formal and final causes in the explanation of nature, according to Miller, reduces humankind to a "horrid life of wiggling impurities, originated in the putrefactive mucus."[10] The study of the Book of Nature would no longer reveal and justify the moral and ethical values that must guide human social behavior. It is only on the view that the First Cause endowed nature with purpose and goal-directedness that the study of the Book of Nature would reward the investigating human mind with greater enlightenment rooted in Divine wisdom: the purpose and goal of the study of the Book of Nature would be to achieve a more profound knowledge of the First Cause, and that includes recognition of the same moral and ethical values in nature that are also laid out in the Book of Revelation.

5.2   Ethics in Nature

Darwin's theory of species transformation based on nothing but natural causes, such as variation and natural selection, might be thought to have ultimately rendered obsolete the search for moral and ethical values in nature. Although not exemplified in nature, these values could, nevertheless, still have their evolutionary roots in nature. Darwin devoted an entire chapter of his *"Descent of Man,"*[11] first published in 1871, to this issue. Darwin argued that morality first arouse in the form of social instincts, which evolved under the influence of natural selection. He saw ample evidence for social instincts in nature, not only in social insects, but also in the parental behavior of birds and mammals. With the development of rational thought, humans are able to reflect on past actions and project future activities subject to moral and ethical considerations. For Darwin, "a moral being is one who is capable of reflecting on his past actions and their motives – of approving some

---

[9]Cited from Secord, J.A. 2000. Victorian Sensation. The Extraordinary Publication, Reception, and Secret of *Vestiges of the Natural History of Creation*, The University of Chicago Press, Chicago, p. 245.

[10]Miller, H. 1849 [1850], Foot-Prints of the Creator: or, the *Asterolepis* of Stromness, Agassiz, L. (Ed.). Gould, Kendall and Lincoln, Boston, p. 337.

[11]Darwin, Ch. 1871. The Descent of Man, and Selection in Relation to Sex. John Murray, London.

and disapproving others."[12] However, to seek justification for moral or ethical values in nature had, long before Darwin, been denounced by the philosopher David Hume[13] as the fallacious attempt to deduce what "ought to be" from what simply "is." Knowledge of what "is" in nature cannot rationally justify what "ought to be" in human society. To bring out the issue in a modern context, lets transpose Hume's argument into a time after the publication of Darwin's "*Origin*" in 1859: on Hume's account, it means to succumb to a logical fallacy if Social Darwinism (i.e., brute and unconstrained capitalism) is justified with reference to Darwin's "theory of natural selection." The converse form of this argument brings out its fallacious nature particularly nicely: the rejection of the doctrine of the "survival of the fittest" by a socially motivated democracy does not imply that Darwin's theory of evolution is false. That theory applies to what simply *is* in nature, and has no bearing on what *ought to be* in society.

Going beyond Hume's argument in his book "*Principia Ethica*" of 1903[14], George E. Moore introduced the concept of "naturalistic fallacy" in his criticism of the attempt to define what is "good" in terms of natural properties. He argued that the term "good" is a primitive term, that is, one that cannot be defined by the use of other terms but that can only be used in the definition of other terms. Darwin thought "as happiness is an essential part of the general good, the great-happiness principle serves as a nearly safe standard of right or wrong."[15] But to define what is "good" in terms of happiness is fallacious, since "good" does not necessarily, i.e., in all contexts, mean to be "happy." Similarly, it is to commit a naturalistic fallacy if the term "good" is defined in social-Darwinist terms such as the "survival of the fittest": values such as "good" cannot be read into nature in an attempt to elucidate their meaning and to find support for their justification. The first lesson learnt from the dissection the metaphor of the two books, then, would seem to be that moral and ethical values cannot be derived from laws of nature. Concerns about moral and ethical values are not related, nor can they be related to a theory of evolution that is built on natural causes and nothing else. The development of such theories stripped nature of inherent moral values and ethical standards, and with those disappeared purposefulness and goal-directedness from nature. As Thomas Huxley, also known as Darwin's bulldog, famously stated in 1894: "The thief and the murderer follow nature just as much as the philantropist."[16] And yet, the current discussion of evolutionary ethics was reignited by Edward O. Wilson's highly innovative, if controversial[17] book "*Sociobiology*," first published in 1975. According to Wilson,

---

[12]Darwin, Ch. 1888. The Descent of Man, 2nd. Ed. John Murray, London, vol. 2, p. 427.

[13]Hume, D. 1740. A Treatise of Human Nature. Reprinted 1978, Clarendon Press, Oxford.

[14]Moore, G.E. 1903. Principia Ethica. Cambridge University Press, Cambridge, UK.

[15]Darwin, 1888, ibid., p. 428.

[16]Huxley, T. H. 1894. Evolution and Ethics, pp. 57–174. In: Paradis, J., and G.C. Williams (Eds.), T.H. Huxley's *Evolution and Ethics*, with New Essays on its Victorian and Sociobiological Context. Princeton University Press, Princeton, NJ.

[17]Segerstrale, U. 1986. Colleagues in conflict: an 'in vivo' analysis of the sociobiology controversy. Biology & Philosophy, 1: 53–87.

"scientists and humanists should consider together the possibility that the time has come for ethics to be removed temporarily from the hands of the philosophers and biologized."[18] The discussion of evolutionary roots of ethics and morality continues, and at the same time the arguments advanced by Hume and Moore remain challenging.[19]

5.3   Universal Laws of Nature

However, the appeal to universal natural laws had further important consequences quite unrelated to issues of ethics and morality, ones that resulted from the fact that science strives to formulate these laws in the language of mathematics. The language of mathematics is timeless, as are also the laws of nature. There *was* a storm yesterday, the sun *will* hopefully shine again tomorrow, but $2 + 2$ *is* four, always, anywhere, at any time. If that is true, as indeed it seems to be, and if it is true that the laws that science seeks to discover are equally timeless, then knowledge of such universal laws of nature would license a particular from of scientific inference called *deduction*. Universal laws can form the basis of deductive inference, a property that was famously exploited by Karl Popper (of whom, again, more later). His simple, intuitively accessible example was a law that says "*All* ravens are black." From it, the prediction "there is no white raven here now" can be deduced, and if the law is true, then that deduction holds always, at all times and everywhere in the universe, quite independently from our observations. If the law is true, there could not be a Leibnizian possible world in which a white raven could be located. Given the truth of the law, the conclusion must be true also, because the conclusion is *logically* entailed by the law. This means that as long as the law holds, the conclusion holds *necessarily* also. Deduction is truth preserving, and truth – in this context – is taken to be timeless. We neither know if the law is true, nor indeed can we know such universal laws to be true: but *if* the law is true, *then* conclusions deduced from it must also be true.

Deduction is a rational mode of reasoning, which, if adopted and correctly performed, allows no dispute over differences of opinion. If it is true that "*all* humans are mortal," and if it is also true that "Socrates is human," then it must necessarily be true that "Socrates is mortal." One might argue about the premises of that conclusion, but once the premises are accepted, the conclusion must necessarily be accepted also. If the premises are true, the conclusion is true, always and everywhere. But notice, the conclusion, and the law on which it is based, does not

---

[18]Wilson, E.O. 1975. Sociobiology, the New Synthesis. The Belknap Press at Harvard University Press, p. 562.

[19]For a brief, useful and accessible introduction to evolutionary ethics, on which part of this account is based, see: http://www.iep.utm.edu/evol-eth/. See also See also Rieppel, O., 1989. Unterwegs zum Anfang. Geschichte und Konsequenzen der Evolutionstheorie. Artemis Verlag, Zürich.

specify when, where, and under which circumstances Socrates would die. It only specifies that Socrates will necessarily die at some point in time and space if he is human. There is no rational dispute possible concerning such deductive inference. Somebody who claims that yes, it is true that all humans are mortal, and yes, it is also true that Socrates is a human, but no, Socrates is for that reason not necessarily mortal, can do so only by slipping into an irrational mode of discourse. Thus, deduction is characterized by two important aspects: it is rational i.e., logical and it is truth preserving. "All swans are white" allows the deduction "there is no black swan here now." Even in the face of black swans this deduction must be considered to be logically valid, yet it would be factually unsound. *If* it is true that "all swans are white," *then* there cannot be any black swan in any possible world. It's not this logical (deductive) inference that is shown to be false by the discovery of black swans. Rather, the discovery of black swans indicates that it is just not the case i.e., it is not true that "all swans are white." Logical relations, such as deductive entailment, hold between sentences and the thoughts they express, not between sentences and the world these describe. The discovery of black swans in Australia does not, therefore, invalidate the deductive inference, but one of its premises instead. It is this logical stringency of deductive reasoning that makes it so attractive for science, since if deductive reasoning could be applied to the world of experience, then statements about this world of experience could – at least potentially – be true or false in the genuine sense of the timeless concepts of truth and falsity.

If universal laws of nature would allow the study of nature in terms of deductive logic, if these laws can appropriately capture nature, then nature would appear to be rationally structured. The "Book of Nature" would thus hold the promise of revealing the universality and rationality of the First Cause. But being bound by universal laws, the "Book of Nature" would reveal the fundamental structure of the universe, rather than describe the particularities of the here and now, for example the particularities of Socrates' death. Although such reasoning may come close to the views held by Kepler and Galileo, they do not quite reflect the views of Karl Popper or other modern philosophers of science. Nature is just the way it is, possibly an enormously complex bundle of intertwined and interacting processes, each one a concatenation of events, and in its historical dimension quite possibly unique.[20] Time obtains from the passing of nature, and through its passing nature acquires a history. But just as it is impossible to reverse the passing of time, so it is impossible to reverse the course of history to start it all over again. The historical processes that are manifest in nature are unique, irreversible, and unrepeatable processes. The universal laws of nature that he appealed to in his philosophy of science were for Popper[21] not inherent in nature but, instead, inherent in the language of science used to talk about nature. It is the language of science,

---

[20]Popper, K.R. 1982 (1998). The Open Universe. An Argument for Indeterminism. Routledge, London, p. 45.

[21]Popper, K.R. 1982 (1998), ibid., p. 45.

especially the laws and theories written in the language of mathematics, which obeys or violates the laws of logic, not nature. These laws may, or may not, successfully apply to nature: if they do, we seem to have come closer to an explanation of natural processes. If, instead, such laws clash with nature, it is not nature that is falsified – it is the laws that scientists talk about that are deemed wrong.

It is easily understood why the notion of universal natural laws first gained a foothold in sciences such as physics and astronomy. If nothing else, it was the regularity and predictability of the movements of celestial bodies, indicating a regularity and uniformity of observable phenomena, which invited the analysis of nature in terms of the time-independent language of mathematics. Days come and go, moons come and go in apparent harmony with the tides, yet timeless mathematical equations were found to almost miraculously capture the ever-changing world with great precision and clarity. Although continuously moving, and hence continuously changing, celestial bodies move on fixed and uniform trajectories through time and space, and these trajectories can successfully be described in terms of mathematical equations. Aristotle took movement to be a fundamental kind of change. A moving object constantly undergoes change relative to its position in space or, in more modern terms, a moving object changes its position in space relative to other objects. The continuous movement of the planets shows the world to be subject to continuous change, but the fact that those movements can be described in terms of universal natural laws was taken to indicate that no real change takes place. The universe that is governed by universal natural laws is one of dynamic permanence. Hence, the power of the metaphor of the two books: the universality and eternity of the fundamental structure of the universe was taken to mirror the universality and eternity of its Creator.

5.4   Lawful Development and the Doctrine of Pre-Existence

What would hold for physics and astronomy was naturally thought to hold for biology also. The search was thus on for a biological understanding of the animate world that would also be one of dynamic permanence, and as was already mentioned, the direction this research was about to take had been pointed out by Aurelius Augustinus. Change seems to take place in nature as an acorn grows into an oak tree, but that only appears to be the case, as the oak tree is already preexistent in the acorn. We have already touched upon the issue of how this doctrine of pre-existence was used by philosophers and naturalists, such as Leibniz and Bonnet, to address the paradoxical problem of change, but it seems worthwhile at this juncture to explore that doctrine in a little more detail to see how exactly the metaphor of the two books was carried into biology. In virtue of their natural beauty, butterflies have attracted human attention at all times. But consider the humble beginnings of such beautiful creatures, the caterpillars. At some point of the caterpillar's development, a metamorphosis takes place: the caterpillar changes into a chrysalis, from which emerges the butterfly, stretching its wings and taking off. Insect metamorphosis, and particularly that of butterflies, motivated strong metaphorical pictures that influenced authors from Ancient times all the way through to Darwin and beyond.

Basilius Magnus, who was later known as St. Basil the Great, was born in Caesarea, the capital of Cappadocia, probably around 329. Cappadocia was part of what is now known as Turkey. After attending the University of Athens, he returned to Caesarea, where he founded a small monastic community. Pursuing a career in the Eastern Church, he eventually was ordained Bishop of Caesarea in 370. From a series of lectures he delivered in Caesarea, he composed his book *"Hexameron"* or *"On the Six Days of the Creation,"* which reveals his deep interest for natural history, and the lessons he had learned in that subject matter by reading Aristotle. In one of these famous lectures, Basilius used insect metamorphosis as a simile for resurrection: "...remember the metamorphosis of the silk-moth which you observe every year: seek a clear understanding of resurrection and do not deny the belief in the transformation which the apostle Paulus has promised us all."[22] So, the belief in transformation is justified, but what is required is a "clear understanding" of such a process of change. Basilius Magnus thus brought into the world a most powerful picture, the influence of which we can trace all the way into Darwin's notebooks.

Jan Swammerdam was a Dutch biologist, born in Amsterdam in 1637. He earned an M.D. at Amsterdam University in 1667, but whereas he did contribute to anatomical studies, he is best known for his work on insects. Since the Netherlands were the home of the earliest microscopes, with Anthony van Leeuwenhoek taking the lead in the development of these instruments, it is no surprise that Swammerdam should make good use of magnifying lenses in his dissections. Swammerdam's interest was particularly drawn to the investigation of insect metamorphosis. He was the first author to conclusively demonstrate that larva, pupa, and adult imago are nothing but different life forms of one and the same (numerically identical) individual organism. His dealing with the paradoxical problem of change created an ambiguity in his writings as to whether he accepted the doctrine of pre-existence, or whether he allowed for genuine transformation in insect metamorphosis. After Swammerdam's death in 1680, his written estate was collected and translated into Latin by the Dutch botanist Herman Boerhaave, and published in 1737 under the title *"Biblia Naturae."* The title, "A Bible of Nature" made a direct appeal to the metaphor of the two books: "There is nothing in the world of nature which deserves greater admiration than the change of a caterpillar to a winged insect."[23] However, at least on one reading, Swammerdam appears not to have thought of insect metamorphosis as a process of true and radical change, and he appears not to

---

[22]Basilius Magnus, 1951. Homélies sur l'Hexaéméron. Texte grex, introduit et traduit par S. Giet. Edition du Cerf, Paris, p. 437.

[23]Swammerdam, J., 1752. Bibel der Natur. Johann Friedrich Gleditschens Buchhandlung, Lepizig, p. 3.

have believed in anything new developing inside the pupa. Instead, he thought that the caterpillar and the butterfly are indeed *essentially* the same creature, yet appearing under different forms at different stages of its life. No essential change took place, nothing really new developed; it was only the outward appearance of the animal that transformed during metamorphosis. Swammerdam believed the butterfly to pre-exist within the structures of the caterpillar. How did he reach that conclusion? Swammerdam studied the molding process of caterpillars in great detail. The "hairy" caterpillars periodically shed the outer keratinized layer of their skin. Inspecting these empty molds, Swammerdam found that the outward projections, which represent the old bristles, were hollow inside. Evidently, the "new" bristles had pre-existed inside the old ones just as the butterfly would pre-exist inside the caterpillar. Influenced by the French philosopher P. Nicholas Malebranche, who had claimed that invisibility is not yet proof of nonexistence, Swammerdam became – or at least was understood as – an early and eminent proponent of the doctrine of pre-existence[24], a theory that held the promise to explain flawlessly a biological world of dynamic permanence: "If this line of thought is carefully considered and appreciated, the error becomes apparent committed by those who want to prove the resurrection of the dead by pointing at the natural and easily understood changes observed in the course of natural events... because, as a matter of fact, those animalcules never die, as man does before his resurrection... yet their metamorphosis is so wonderful that one could easily be misled to believe that a new animal was born and emerged anew from the previous condition of existence. This is the limit to which insect metamorphosis can be stretched as a simile of resurrection."[25] This quotation from Swammerdam's "*Biblia Naturae*" introduces insect metamorphosis not only as an example of biological dynamic permanence, but also uses this phenomenon as a reference point for the discussion of the dogma of resurrection. In a later chapter of this treatise, Swammerdam elaborates on his "comparison of man with insects and frogs," "Just as the caterpillar seems to lack wings, so does the tadpole seem to lack limbs, but again corresponding to the metamorphosing insect, the limbs of frogs can be seen to develop from pre-existing limb-buds."[26] Chambers and his opponent Miller were not the first to think of it: embryology vaguely recapitulated the Great Chain of Being in Swammerdam's world of ideas already. The tadpole corresponds to the fish stage, recapitulated by frogs, a group of tetrapods.

The German philosopher Gottfried Wilhelm Leibniz synthesized contemporary biological thought in the preface to his *Theodicy* of 1710, admitting that he shared the greatest admiration for the discoveries of his predecessors, Swammerdam being one of them, and the other – as already discussed – Antony van Leeuwenhoek. Leibniz was hailing the doctrine of pre-existence as a means to support the view

---

[24]Wilkie, J.S. 1967. Preformation and epigenesis: a new historical treatment. History of Science, 6: 138–150. See also Rieppel, 1989, ibid.

[25]Swammerdam, 1752, ibid., p. 9.

[26]Swammerdam, 1752, ibid., p. 313.

that the natural course of events was regulated by secondary causes enacted by the First Cause, thus keeping the Creator outside the constraints of time and space: "...if God is not believed to form organic beings by means of continuous miracles, the conclusion becomes inevitable that the Creator preformed things in a way that new forms of organization are nothing but the physical consequences of a preceding form of life, just as butterflies evolve from caterpillars as a mere consequence of an unfolding of pre-existent structures, as Mr. Swammerdam has convincingly shown."[27] The only kind of "evolution" Leibniz was prepared to accept was the "unfolding" of pre-existent structures, similar to the unfolding of a flower bud. This is the fundamental difference between Leibniz's world of natural phenomena, regulated by time-independent natural laws, which would not allow the origin of anything genuinely new, and the natural world of Chambers, which would permit the emergence of new structures characterized by new properties.

Charles Bonnet, the quintessential eighteenth century *naturaliste philosophe* from Geneva, whose discoveries and ideas were already discussed in earlier chapters, admitted the profound influence the reading of Leibniz's *"Theodicy"* exerted on him during the winter of 1748.[28] Independent of Leibniz and before reading his *"Theodicy,"* Bonnet had adopted the doctrine of pre-existence. In Bonnet's mind once again, the embryonic unfolding of an organic being recapitulated the succession of inferior modes of organization. This was not only reflected in the similarity of the early appearance of a chicken embryo to a maggot, as Bonnet believed, but even more so by the fact that the behavior of small children corresponded in his view less to the well reasoned behavior of adults than to the playful behavior of higher animals. According to Bonnet, "children acquire the status of rational beings through the development of all their attributes and through education... whereas today, animals are in a condition of childhood; perhaps, one day, animals will develop into rational beings!"[29] However, "development" was used in a Leibnizian sense in this context, meaning the unfolding or actualization through development, of a predetermined and pre-existent potential.

5.5 Universal Laws and Genuine Change

Pushing the metaphor of the two books that far, a crossroad had been reached, where ideologies had to part: dynamic permanence vs. true change in time. Erasmus Darwin, Charles' illustrious grandfather, forged an evolutionary world-view in his *"Zoönomia,"* the first volume of which was published in 1794. There, he pondered

---

[27]Buchenau, A. (Ed.), 1968. Gottfried Wilhelm Leibniz: Die Theodizee. Felix Meiner, Hamburg, p. 23.

[28]Rieppel, O. 1988. The reception of Leibniz' philosophy in the writings of Charles Bonnet (1720–1793). Journal of the History of Biology 21: 119–145.

[29]Bonnet, C. 1769. La Palingénésie Philosophique, Vol. 1. C. Philibert & B. Chirol, Genf, p. 317.

"the great changes, which we see naturally produced in animals after their nativity, as in the production of the butterfly with painted wings from the crawling caterpillar; or the respiring frog from the subnatant tadpole... when we consider all these changes of animal form, and innumerable others... we cannot but be convinced, that the fetus or embryon is formed by apposition of new parts, and not by the distention of a primordial nest of germs, included one within the other, like the cups of a conjurer."[30] The rejection of theories of pre-existence, and the adoption of an atomistic conception of animal generation, opened the door to evolutionary transformation. Recapitulating his findings of animal generation, Erasmus Darwin concluded: "At the nativity of the child it deposits the placenta or gills, and by expanding its lungs acquires more plentiful oxygenation from the currents of air... like the tadpole, when it changes into a frog, becomes an aerial animal... so from the beginning of the existence of this terraqueous globe, the animals, which inhabit it, have constantly improved, and are still in state of progressive improvement... This idea of the gradual generation of all things seems to have been as familiar to the ancient philosophers as to the modern ones."[31] All of this generating nature was tied into continuity of cause and effect: "Cause and effect may be considered as the progression, or successive motions, of the parts of the great system of Nature. The state of things at this moment is the effect of the state of things, which existed in the preceding moment; and the cause of the state of things, which shall exist in the next moment."[32] In his early notebooks Charles Darwin mused: "There is an analogy between caterpillars with respect to moths & monkeys & man – each man passes through its caterpillar stage. The monkey represents this state."[33] It is well known that Darwin compared the behavior of his children when they were little with that of the orangutans named Jenny in the Regent's Park Zoo of London. Of Jenny, Darwin noted in his 1838 notebook: "Let man visit the ouran-outang in domestication... Man in his arrogance thinks himself a great work... More humble and I believe true to consider him created from animals."[34] These were precisely the conclusions Miller had anticipated to emerge from reading Lamarck's, and later Chambers' book, and he was determined to preempt such false inferences. The old metaphors and images again emerged in his writings: "The Egyptians wrapped up the bodies of their dead in the chrysalis form, so that a mummy, in their apprehension, was simply a human pupa, waiting the period of its enlargement; and the Greeks had but one word in their language for butterfly and the soul. But not the less true is it, notwithstanding, that the facts of insect transformation furnish no legitimate key to the totally distinct facts of a resurrection

---

[30]Darwin, E. 1794. Zoönomia, or, the Laws of Organic Life, vol. 1. J. Johnson, London, sect. XXXIX, IV, § 8.

[31]Darwin, E., 1794, ibid., sect. XXXIX, VII, § 8, 9.

[32]Darwin, E., 1794, ibid., sect. XXXIX, VIII, § 1.

[33]DeBeer, G. 1960. Darwin's notebooks on transmutation of species, part III. Bulletin of the British Museum (Natural History), Historical, 2:148.

[34]DeBeer, G. 1960. Darwin's notebooks on transmutation of species, part II. Bulletin of the British Museum (Natural History), Historical, 2:106.

of the body, and of a life after death."[35] The metaphor, for him, had been pushed too far!

However, if – as Chambers believed – the earth was to have emerged from the initial chaos, if life on earth would have evolved toward ever increasing complexity, if – in short – earth history would have been progressive, then the eternal cycle of uniform movement would have to be broken up in favor of a directional history with a beginning in time. If the earth and its biota have a unique history with a beginning in time, then an account of that history in terms of eternal and universal laws of logic, or of equally timeless and universal laws of nature, could potentially become problematic. It is difficult to say that "2 + 2 = 4" should not have been true at any one time or place in the past, but it is not difficult to say that Miller's *Asterolepis* existed in the past but no longer exists today and will never exist again in the future. It seems worthwhile, then, to consider the applicability of the notion of universal natural laws in a science that seeks to explain an evolving world subject to irreversible change through time.

It was once again the philosopher Gottfried Wilhelm Leibniz who was among the first to propose a process of cosmic evolution. In his *"Protogaea,"* published in 1749, he argued "initially the earth was uniform and fluid, since God would not create anything disproportionate. Solid parts have originated from fluids."[36] As was already discussed, the eighteenth Century French naturalist Georges Buffon expanded this view into a theory of a cooling earth.[37] The immortal German philosopher Immanuel Kant from Königsberg, now Kaliningrad in Russia, was another protagonist of an evolving universe. In 1755, when only 31-years old and not yet widely known, Kant anonymously published a little booklet on the origin of the universe in accordance with Newton's Laws. In this book, he proposed his "nebular hypothesis,"[38] according to which the planetary system would have evolved from primordial cosmic matter through the action of gravity. Kant thus laid the ground for what historians of science would later refer to as the "Kant-Laplacian theory." Pierre-Simon Marquis de Laplace (1749–1827), whom Chambers cited in his *"Vestiges,"* abandoned the career track that had been laid out for him by his father when he developed his mathematical skills and interests while studying theology at the University of Caen in Normandy, France. His subsequent work in mathematics and, later, astronomy laid the foundation of what would come to be known as "mathematical astronomy," a research program that was to set new standards for natural sciences, biology included.[39] In his treatise entitled

---

[35]Miller, 1849 [1850], ibid., p. 178

[36]Scheid, C. (Ed.), 1749. Protogaea, by Gottfried Wilhelm Leibniz. Johann Gottlieb Vierling, Leipzig, p. 39.

[37]For details and references see Rieppel, O. 1989. Unterwegs zum Anfang; Geschichte und Konsequenzen der Evolutionstheorie. Artemis Verlag, Zürich.

[38]Bowler, P.J. 1984. Evolution. The History of an Idea. The University of California Press, Berkley, p. 32.

[39]Winsor, M.P. 1976. Starfish, Jellyfish, and the Order of Life. Yale University Press, New haven, p. 140.

"Exposition du Système du Monde" (*System of the World*), which appeared in 1769 written in a language that omitted technical details, he further elaborated on Kant's original ideas. According to Laplace, the solar system would have evolved as a result of gravitational forces from a rotating cloud of interstellar gas and dust; the sun would have condensed first, followed by the condensation of a system of planets around it. It was a historical coincidence that the telescope constructed by William Herschel, father of John Herschel, the contemporary of Charles Darwin, would draw the condensation of a central star from interstellar clouds of primordial matter closer to vision.[40] The crucial point of all these cosmological models was that the earth, indeed the solar system, was viewed as having gradually evolved from chaos. But this cosmic evolution was governed by eternal and uniformly acting laws of nature, which shaped equally eternal and uniform primordial matter into the universe as we know it. The impact of Newtonian physics on the thinking of these scientists and philosophers is easily appreciated, since all of them considered the Law of Gravity the central guiding principle of cosmology. Laplace was an ambitious man, not too modest to approach the emperor Napoleon Bonaparte in his quest for political honors and employment at the highest levels of government.[41] As a reference on his own behalf, Laplace offered Napoleon a copy of his *"System of the World."* Bonaparte was captured by the power of imagination expressed in Laplace's treatise, but nevertheless found it startling that the Creator was not once referred to throughout the entire book. Calling on its author to account for this omission, Laplace famously answered: *"Je n'avais pas besoin de cette hypothèse-là"* ("I had no need of that hypothesis") – secondary causes were all that was required to do the job of explanation of how our solar system came into being. The rejection by Laplace of the hypothesis of a Creator must be understood strictly in the context of his pursuit of hard science. In his view, such a hypothesis would all too easily, but also trivially, explain everything, but it would not issue testable predictions, unless the Almighty would be content to rest His powers. In contrast, universal natural laws do allow testable predictions, even if of a statistical nature only, and this is what Laplace's science was all about.

If nature was ruled by uniform and universal laws, the natural course of events would be completely determined, the continuous chain of cause and effect entirely predictable. Accordingly, Laplace posed a demon[42] who knows all the lawful causal connections that exist in nature. If such a demon were also in possession of knowledge of all the relevant details of the state that the world would be in at present, he would be able to infallibly predict the future natural course of events through all time. Knowledge of all universal natural laws would equal potential omniscience. Given its present state, the future state of the universe could be

---

[40]Bowler, 1984, ibid, p. 33.

[41]For more details see Fox, R., C.C. Gillespie, and I. Grattan-Guiness. 1978. Laplace, Pierre-Simon, Marquis de, pp. 273–403. In: Gillespie, C.C. (Ed.), Dictionary of Scientific Biography, Vol. 15, Suppl. I. Charles Scribner's Sons, New York.

[42]Schneider. L. 1988, Isaac Newton. Beck Verlag, München, p. 158.

predicted in all detail. It was left to modern physics to refute the hypothesis of a Laplacian demon, i.e., the hypothesis of a complete and universal determination of the world, but at the time of Chambers' or Miller's writing, this thesis still had a major scientific as well as sociological significance. Yet there remained an important difference, in the eyes of critics, between the theories of Laplace, Herschel, and others, who postulated an evolving universe on the basis of universal natural laws of nature and the writing of Robert Chambers who did not restrict the thesis of lawful evolution to dead matter, but expanded it to account for the successive development of life on earth. This, indeed, was perceived as a "conflation"[43] of two fundamentally different "ways of seeing," claimed to be manifest in Chambers' play on words. In his book, he seemed to be using the same words to describe the animate and inanimate world, but almost imperceptibly to the lay public, he changed their meaning in a subtle but most important way. The criticism was that Chambers' notion of a "natural law" that governs the animate world was not the same as the notion of "natural law" that was employed by Newton, Kant, and Laplace in their study of the inanimate world. Newton' and Laplace's laws were explanatory: they link causes with effects – same cause, same effect, always, everywhere. Form there results their predictive power. In contrast, Chambers' "Law of Development" captures the recognition of some putative pattern of order in nature without predictive power.[44] The law merely states the order that prevails in nature in terms of the Great Chain of Being, which is an order that pervades embryonic development, the Fossil Record, and animal classification, but it offers no natural causal explanation of *how* that particular order in nature came about.

5.6   Description vs. Explanatory Laws

Let us take a closer look at Chambers' notion of lawfulness. The foundation on which Kant and Laplace built their cosmology was Newton's Law of Gravity – and it is to these giants that Chambers looked back in support of his own theory. The Law of Gravity explained very successfully the regularity of phenomena observed at nightfall. This was true irrespective of the fact that nobody, not even Newton, could tell what the essence, the nature of gravity was, what the material basis was for this mysterious "attractive force," as it used to be called. The paradigm of Cartesian physics required that extended bodies such as billiard balls hit each other to transmit any "impetus," or motion. But gravity was exerting attraction "at a distance." How could this be possible? Many at the time referred to Newton's gravitational force as a "mysterious force" or an "occult force." Indeed, some concluded with the philosopher David Hume that the notion of such "forces" should

---

[43]Hodge, M.J.S. 1972, The universal gestation of nature: Chambers' *Vestiges* and *Explanations*. Journal of the History of Biology, 5: 132.

[44]Secord, 2000, ibid., p. 407.

be abandoned in any sound philosophical system of nature. Unintelligible as they were, the invocation of such "forces" contributed nothing to a proper understanding of nature. All that laws of nature could be about was the seemingly universal regularity of certain natural phenomena. Natural laws were to capture the regularity of phenomena apparent in nature, whatever their mysterious cause would be.

Today we know that Newton's invocation of an aethereal substance was a mistaken move in his attempt to explain the nature of gravity, but his theory still remains eminently successful in explaining why celestial bodies keep on moving on permanent paths rather than falling from the sky. The success of the Law of Gravity was rooted not in the knowledge of the material basis of this force, but in its explanatory power instead; one could measure the correctness of predictions based on it, and they simply proved nearly perfect. This, indeed, seemed to many to be the crucial issue in scientific investigation: an observed regularity of phenomena is suggestive of underlying lawfulness. Knowledge of the underlying causes would in turn enable scientists to successfully predict future regularity. Many philosophers and scientists argue that the explanatory power of a scientific theory is to be measured by testing predictions derived from it. But for this to be possible, a scientific theory must address recurrent events: no testability without predictability and no predictability without repeatability! Everybody can test the Law of Gravity, simply by picking up a stone or any other object and let it fall back to earth again. If a stone would remain suspended in the air, the Law of Gravity would be partially refuted, unless hallucinations could be invoked. The Law of Gravity would also be falsified if the solar system were to fall apart. Until now, the law has stood the test of time.

Newton's Law of Gravity proved to be the basis for highly successful predictions – the question is whether the Law of Development proposed by Chambers would be as successful in the prediction of natural processes in animated nature? Chambers likewise pretended to derive his law from observed regularity of phenomena, namely the regular succession of fossils through times, which parallels not only the Great Chain of Being, but also the embryonic development of organisms aligned along the ladder of life. His Law of Development was designed to explain the regularity of observation that fossil fishes are consistently found in earlier strata than, let us say, fossil mammals. A similar unbroken regularity pertains to the occurrence of fossil reptiles before humans. And it seemed to him – as it still does to us – that this fact was at least as well established as the fact that a stone will always fall back to earth when picked up. Indeed, nobody has ever found a fossil mammal predating the earliest known fossil fish, or a fossil human predating, or contemporaneous with, a (non-avian) dinosaur.

As was forcefully argued by the British philosopher William Whewell (1794–1866), the strength of support for a scientific theory generally increases with the "consiliense" i.e., the coming together of various purportedly independent lines of evidence. Or in other words, a scientific theory is said to gain in strength when more seemingly disparate observations can be brought under its explanatory umbrella. The Fossil Record, according to Chambers, laid out a certain order in nature, progressing form early simplicity to later complexity. The lawful nature of such progression was re-enforced, in Chambers' view, by the "coming together" of independent evidence documenting the same progressive order, namely the Great Chain of Being, and embryonic development. Johann Friedrich Meckel the Younger (1771–1833), who was destined to become one of the greatest anatomists of the eighteenth century, left Halle an der Saale in 1903 to pursue his studies in Paris under the tutelage of the great Georges Cuvier and Etienne Geoffroy Saint-Hilaire. At the time, both these towering scientists were also collaborating with Etienne Serres (1786–1868), a French physicist and embryologist. Combining Meckel's interests in teratology, the study of malformations, with Serres' interests in embryology in this intellectual melting pot that was the Paris Natural History Museum, the two researches came up with the theory known today as the "Meckel–Serres Law": a twofold parallelism between embryonic development and the Great Chain of Being. To explain this parallelism, Meckel and Serres invoked a theoretical "developmental force." To this concept, other authors, such as Louis Agassiz or Robert Chambers, added the Fossil Record as another line of evidence, such that the Meckel–Serres Law became the famous "threefold parallelism" of later authors. As we have seen, Chambers must have had this "Meckel–Serres Law"[45] in mind when he formulated his Law of Development to explain the temporal succession of the appearance of fossils, and he must have felt justified to treat the causal mechanisms of his law, this mysterious "developmental force", as much as a theoretical, i.e., unobservable entity as Newton had treated the material basis of his Law of Gravity, his "aetheral" substance. So where is the difference?

The difference is that irrespective of their unknown material basis, Newton's Law of Gravity allows testable predictions, but Chambers' Law of Development does not. Surely, the empirical observations pertaining to animal classification, embryology, and the Fossil Record could be subsumed under a "Law of Development," and this law could be claimed to provide a hypothetical explanation of the "threefold parallelism" – but all of that theorizing still did not offer a platform for testable predictions. What testable predictions can be derived from Chambers' Law of Development? None come to mind. William Whewell stripped down the *"Vestiges"* to "a system of order,"[46] an expression of progression from dead matter to plants, from animal to human. Adam Sedgwick accused the author of *"Vestiges"* of having "no experience of the hard work of inductive science."[47] The "hard work" consisted in collecting enough observational evidence from nature that would allow the inference, from that evidence, to laws that were explanatory in a special sense, i.e., not only merely descriptive, but also predictive. Inductive science is not tied to the logical rigor of deduction. Deduction, as was discussed earlier, is rational and truth preserving. It starts from premises, from which conclusions are deduced

---

[45]Gould, S.J., 1977. Ontogeny and Phylogeny. Harvard University Press, Cambridge, MA.

[46]Cited after Secord, 2000, ibid., p. 229.

[47]Cited after Secord, J.A. 1994. Introduction. In: Secord, J.A. (Ed.), Robert Chambers: Vestiges of the Natural History of Creation, and Other Evolutionary Writings. The University of Chicago Press, Chicago, p. xxxi.

according to the laws of logic. If the premises are true, then the conclusion must be true also, because the conclusion is logically entailed by its premises. As Popper has shown, it is possible to simply invent a law of nature, then deduce predictions from it, and subsequently to try to find out whether these predictions apply to the observable world. If they do not, the law is rejected. Inductive sciences are a much more messy enterprise. Here, evidence is collected from the observation of regularity in nature. If an investigator thinks that enough evidence has accumulated, he/she may venture to infer from that observed regularity some underlying lawfulness. But to test the validity of such inference, the inferred law must allow testable predictions.

Many authors – Popper among them – have pointed to several problems with such an analysis of scientific discovery in terms of inductive inference. Some have called it circular: how is it possible, in a noncircular way, to proceed by inference from observed regularity to the prediction of more of the same regularity? Well, again, inductive sciences are somewhat of a messy enterprise, and although such inference of natural laws might be attacked from a strictly logical point of view, it seems to work none-the-less. We do it all the time in our everyday life, and very successfully so. Why should we not also do it in science, especially if sciences prove to be successful in the causal explanation of nature? But right at this juncture obtains a second important critique of inductive sciences: what is a successful explanation of nature through science? The general answer is that such successful explanation of natural processes must link causes to effects, and on that basis allow to make predictions. A regularity of natural processes obtains from the same cause producing the same effect. But that means that if such lawfulness is inferred from an observed regularity of natural events, then the assumption must be made that these events are, indeed, linked to the "same cause–same effect" principle. Lawful regularity in nature requires such regularities to be connected, the same cause connecting with the same effect everywhere, forever, as is required by universal laws of nature. But how could we know? We will never know how far the chain of observed regularity stretches into infinity. We will always remain restricted to the space–time region of our observational experience. So who is to say that the regularity we observe in our observationally accessible space–time region is not purely accidental. Who is to say that events, which appear regular to us, have not just been tossed into our experienced space–time region by some cosmic accident, without any cause relating lawfully to any effect? If we cannot answer this question, how can we then appeal to universal laws of nature in the attempted explanation of the experienced world? How can we expect any prediction derived from putative laws of nature to turn out to be correct on any basis other than pure chance?

So how can the distinction be drawn between causal connectedness and merely accidental regularity? Uniformity of law guarantees uniformity of nature. On that basis, past regularity of the sun rising every morning seems to license the conclusion that the sun will rise again tomorrow. But what licenses the belief in uniformity of nature? Or in other words: which is first, the recognition of past uniformity of nature that licenses the inference to uniformity of law; or the recognition of uniformity of law that licenses the prediction of future uniformity of nature? This is a seriously tricky question that has no easy answer. As we have seen in the previous chapter, some philosophers of science start from laws. For Popper, a scientist conjectures a universal law, or invokes it on the basis of her intuitions, and then proceeds to test it deductively: "All ravens are black – there is no white raven here now." Universality is a property of the stipulated law, not of the world. The consequence is that we will never know such laws to be true, but we can know them to be false. However, and in contrast to Popper, most practicing scientists would probably argue that scientific investigation does in fact start with the recognition of some regularity in nature, which science then sets out to explain. But if that is so, we are left to distinguish between two different kinds of regularities that may prevail in nature. On theoretical grounds, we can imagine regularities in nature that are not governed by natural laws, but that are purely accidental, or "contingent" as philosophers say, as opposed to those regularities that are causally conditioned. After all, scientific investigations always remain limited by time, space, and resources. The arm of science does not extend into eternity and infinity. If it did, universal laws could, at least in principle, be known to be true.

The movements of the stars on the nightly sky seem to be governed by universal laws of nature. The prediction that the sun will rise tomorrow is not just based on the past experience of some accidental regularity, but is also licensed by Newton's Laws.[48] Newton provided an explanatory scientific theory that allows the prediction that the sun will rise again tomorrow. And this same scientific theory does not just explain why the sun rises every day, it explains much more: it integrates a whole host of cosmic phenomena into a unified causal explanation, one furthermore that in a later incarnation turned out to be reliable enough to successfully land people on the moon, or robots on Mars. This is how scientific theories gain strength through consilience. Were it the case that Newton's Laws described merely accidental regularities, rather than explain their causal connectedness, the landing of people on the moon, or robots on Mars, would amount to nothing less but miracles. But miracles are not the stuff from which science is made. Chambers sought similar integrative explanatory power for his "Law of Development," arguing that it explains the perceived parallelism between classification along an ascending gradient of complexity, embryonic development, and the Fossil Record. Although the earth continues to orbit the sun, the universe originated only once, just as life on earth is believed to have originated only once. The ascent from fish to human happened only once, none of the innumerable steps of transformation that link humans to fish is recurrent, repeatable, or in its entirety reversible. Chambers' evolutionism invoked a unique historical process of transformation to explain the past and present ascent of life to ever-increasing complexity – but a unique historical process cannot possibly serve as basis for the inference of universal laws from which testable predictions can be deduced. History does not, or at least it needs not repeat itself. Sciences such as physics and astronomy seem to differ

---

[48]Sober, E. 1988. Reconstructing the Past. Parsimony, Evolution, and Inference. The MIT Press, Cambridge, MA.

fundamentally from historical sciences, because they address natural processes uniform and reproducible enough so as to be captured by mathematical equations. Historical sciences, on the contrary, investigate a unique concatenation of irreproducible events as they unfold through time, a process that remains essentially unpredictable: no testability without repeatability. It is possible to model historical developments in a probabilistic fashion, but it is not possible to deduce such models from past history. It is for this reason that the so-called "hard" sciences such as physics, or astronomy, have been called explanatory sciences, whereas historical sciences have been called as "descriptive" or "narrative." Physics is concerned with the explanation of natural processes in terms of underlying causality, and historical sciences have been characterized as being concerned with the detailed description of unique historical situations or events. This discrepancy between experimental and historical sciences motivated Henry Gee, senior editor for the prestigious scientific journal "*Nature*," to trace the ways by which modern paleontologists went about to carry their science beyond a mere historical narrative into the realm of testability, because "no science can ever be historical."[49] On that account, if evolutionary theory were strictly and only a narrative of the history of life on earth, it would not be scientific. Henry Gee here followed a popular interpretation of Popper's philosophy of science, one that was also sketched by a prominent contemporaneous systematist and vertebrate paleontologist[50] with whom he interacted. It was, indeed, an understanding of science that was to some degree promoted by Popper himself. Popper identified as scientific those theories that are testable and at least potentially falsifiable. For Popper, theories that fulfill this requirement had to take the form of universal laws, from which predictions could be deduced. Evolutionary theory, concerned as it is with the unique history of life on planet earth, cannot take this form. Consequently, for Popper, the total edifice of evolutionary theory was not true science. Rather, it was for him a "metaphysical research program,"[51] composed in its totality of an array of truly scientific subdisciplines such as genetics, physiology, etc. Popper later tried to prevent further misunderstandings by clarifying his position on evolutionary theory.[52] However, judging Newton's Law of Gravity against Chambers' Law of Development from this perspective shows that Chambers' problem was not just the distinction between causally determined as opposed to merely accidental regularity. It was the fact that Newton was working in "hard science," one that satisfies the criterion of testability, which just *is* the criterion required to distinguish causally determined as opposed to accidental regularity. Chambers, in contrast, was concerned with history, which led him to the description of an ordering principle that runs through the classification of

---

[49]Gee, H. 1999. In Search of Deep Time. Beyond the Fossil Record to a New History of Life. Free Press, New York, p. 8.

[50]Patterson, C. 1978. Evolution. British Museum (Natural History), London, p. 149.

[51]Popper, K.R. 1974. Autobiography, pp. 1–181. In Schilpp, P.A. (Ed.), The Philosophy of Karl Popper, vol. 1. Open Court, La Salle, IL.

[52]Popper, K.R. 1980. Evolution. New Scientist, 21 August 190: 611.

organisms as Chambers saw it, their embryonic development, and the Fossil Record.[53] But to describe some succession of events is not also a causal explanation of that succession of events, nor does it provide any basis for the testable prediction of future events.

5.7 Darwin's Law of the "Higgledy-Higgledy"

Darwin naturally faced the same difficulties that arise when historical sciences are pitched against physics. He obviously had recognized the weakness of Chambers' argument. With the clairvoyance characteristic of the *"Origin of Species"* (1859) he wrote: "The author of the 'Vestiges of Creation' would, I presume, say that, after a certain unknown number of generations, some bird had given birth to a wood-pecker, and some plant to the mistletoe, and that these had been produced perfect as we see them; but this assumption seems to me to be no explanation, for it leaves the case of the coadaptations of organic beings to each other and to their physical conditions of life, untouched and unexplained."[54] Yet, he, too, aspired to bring historical biology up to the stringent standards characteristic of the paradigmatic sciences of physics and astronomy, both written in the language of mathematics. He, too, spoke of "laws of nature" such as the *"Law of Variation,"*[55] and during the early phases of his theorizing he considered the principle of natural selection to be of an equivalent status[56] (i.e., to be a secondary cause enacted by the First Cause for the natural creation of new species). The "metaphor of the two books" becomes once again apparent, this time in Darwin's early musings on the transformation of species, although he would later abandon it. As was explained earlier, the attraction of that metaphor was rooted in the fact that timeless mathematical equations could perfectly capture the continuous motions of heavenly bodies. The astronomer Sir John Frederick William Herschel, following in the footsteps of his famous father in the search for universal laws of nature, claimed that "all [my] endeavors have a common feature: they can be interpreted as an attempt to *annihilate* time, as a search for the constant amidst change."[57] He, like so many others mentioned earlier, saw the world as one of the dynamic permanence, ruled by time-independent lawfulness.[58] This is a stark contrast to Darwin's point of view, who looked at the

---

[53]Rudwick, M.J.S. 1972. The Meaning of Fossils. Macdonald, London, p. 226. Gillespie, 1979, ibid., p. 36.

[54]Egerton, F.N. 1970. Refutation and conjecture: Darwin's response to Sedgwick's attack on Chambers. Studies in the History and Philosophy of Science, 1: 178.

[55]DeBeer, G. (Ed.) 1960. Darwin's notebooks on transmutation of species, part III. Bulletin of the British Museum (Natural History), Historical, 2:141

[56]Ospovat, D., 1981, The Development of Darwin's Theory. Natural History, Natural Theology & Natural Selection 1838–1859. Cambridge University Press, Cambridge.

[57]Cited from Schweber, 1989, ibid., p. 34.

[58]Schweber, 1989, ibid., p. 42.

world as one of continuous historical change. It is obvious that any unique historical process cannot possibly be captured in terms of time-independent, universal laws of nature. Compare Herschel's understanding of "proper science" with what Darwin had to say about his "Law of Natural Selection": "It may be said that natural selection is daily and hourly scrutinizing, throughout the world, every variation, even the slightest; rejecting that which is bad and adding up all that is good; silently and insensibly working, whenever and wherever opportunity offers, at the improvement of each organic being in relation to its organic and inorganic conditions of life. We see nothing but slow changes in progress until the hand of time has marked the long lapse of ages, and then so imperfect is our view into the long past geological ages that we only see that the forms of life are now different from what they formerly were."[59] Here, time figures prominently, the driving force of natural selection working daily and hourly as well as over long past geological ages – but even worse. Darwin admitted to the "imperfectness of our view," our vision becoming blurred by the passage of time. For some of his critics, the ultimate mark of bad science, though, must have been Darwin's appeal to *opportunities* that may or may not obtain for natural selection to do its work. This is a bold appeal to *accidents* as an explanation for evolutionary change.[60] Small wonder, therefore, that Herschel, who had adopted from his father the logical stringency of scientific reasoning and a rigorous notion of what constitutes a natural law, could only belittle Darwin's attempt to bring historical biology to live up to the standards of physics and astronomy. In his classic, the three-volume *"Principles of Geology,"* published between 1830 and 1833, Charles Lyell had commented on the succession of different species through geological time, but while he acknowledged the extinction of species, he left open the question how these were replaced in time by different species. In a letter to Lyell from February 20, 1836, John Herschel had called the "law of life"[61] that would explain the replacement of species through geological time the "mystery of mysteries." In the opening pages to his *"Origin"* (1859), Darwin promised "to throw some light on the origin of the species – that mystery of mysteries, as it has been called by one of our greatest philosophers."[62] Could his principle of natural selection be that mysterious "law of life"? – "The law of the higgledy-piggledy," Herschel called it.

This may have been well-calculated polemics, but it did little justice to Darwin's attempts to formulate a comprehensive and integrative theory that would causally explain the historical dimension of the biodiversity found on earth rather than just describe it. What exactly was wrong, or at least arrogant, in Herschel's dismissal of natural selection as the "law of the higgledy-piggledy"? It was his overly stringent notion of what "proper science" – bound as it were by universal laws of nature – would

---

[59]Darwin, Ch, 1859. On the Origin of Species. John Murray, London, p. 84.

[60]Nyhart, L.K. 1995. Biology Takes Form. Animal Morphology and the German Universityies, 1800 – 1900. The University of Chicago Press, Chicago, p. 109.

[61]Secord, 2000, ibid., p. 92.

[62]Darwin, 1859, ibid., p. 1.

have to be. Or, to put it differently, by delivering a scientific explanation for the origin of species through variation and natural selection, Darwin had to abandon the aspiration to bring biology up to the standards by which physicists and astronomers of his time measured the merits of scientific explanations. He had to give up the notion of universality of natural laws, since this notion cannot gain purchase in historical sciences. The explanations that Darwin invoked were constrained by space and time and hence no longer universal. Necessity, and with it certainty, had to give way to probability.

Unlike Chambers, Darwin would not rest content with a mere description and classification of biodiversity, past and present. He wanted to seek a scientific, that is, a causal explanation for the origin and diversification of species, and Charles Darwin was, indeed, the first to achieve that goal. However, the pattern of explanation that Darwin introduced with his theory of evolution differed from the pattern of explanation that at his time was claimed to characterize physics and astronomy, the king and queen of natural sciences. With his theory of evolution, Darwin not only provided a successful causal explanation of the origin of new species, but also validated a different way of doing science. Or at least he validated a way of doing science that was considered illegitimate by the physicists and astronomers of his time, but that was highly successful in historical biology. This is where Chambers failed, and Darwin triumphed.

First off, and rather trivially: the arm of science does not reach into infinity. Science is partitioned in to disciplines and subdisciplines, and scientists are specialists in one or another branch of the special sciences. Different specialists talk about different areas of interest, their theories therefore range over different domains of discourse. Plant physiologists may take an interest in photosynthesis; animal physiologists may take an interest in digestion. The theory of photosynthesis ranges over green plants, but neither over fungi, nor over animals. The interplay of gastric acid and digestive enzymes that break down the ingested food in the stomach of animals is not something botanists normally talk about, unless perhaps at a symposium dinner where one participant suffers from hurtful heartburn. Laws of nature as science knows them come with a certain scope, they generalize, or explain, over a certain domain of interest, where that domain can be broad or narrow. Popper characterized universal laws of nature as being of the form of "*All*-sentences": "All ravens are black." Correspondingly, "All fishes are infected with XYZ" is a universal statement that cannot be known to be true for reasons discussed earlier. "All fishes in Lake Baikal are infected with XYZ" is a statement that generalizes over a broader domain of discourse than "All fishes in the pond in my backyard are infected with XYZ," but both these latter statements can at least potentially (i.e., in principle, even if difficult in practice) be known to be approximately and relevantly true, such that countermeasures against the infection would seem justifiable or warranted. When talking of natural laws, or scientific theories, one must therefore always keep in focus the domain of discourse within which a law, or theory, is supposed to hold. The theory of gravity is of a much broader scope than the theory of natural selection. Everything that is subject to natural selection is also subject to gravity, but not everything that is subject to gravity is also subject to natural selection – Mount Everest for example is not. So care is to be taken if theories of physics are compared with those of biology.

Second, if the old empiricists were right, and scientific investigation does indeed start with observation, no matter how theory-laden that may be, it would have to be the regularity of recurrent observations, or recurrent events, that would cry out for explanation. A law, or theory, invoked to explain such regularity of observations, or events, would require these observations, or events, to be causally grounded. A mere accidental, or fortuitous, regularity of events will neither allow a causal explanation in terms of a natural law that is successful in predicting future observations nor will it allow the prediction of events that could be used to test that law or theory. Here again, it must be noted that a *description* of regular patterns in nature is, indeed, a first scientific achievement, or rather can be one, but it is not also an *explanatory* achievement.[63] It is, after all, the possibility to derive from a theory testable predictions that allows the distinction of causally grounded vs. accidental regularity.

However, the idea that natural laws, or scientific theories, must be grounded in causal relations that govern regularly recurrent events creates serious problem for the historian, because regularly recurrent events are rare in history. How, then, is the causal grounding of theories possible in historical research, which seeks to explain singular events, such as the origin of a new species of plant or animal? There must be a way to distinguish singular events that are just accidental, fortuitous, and freakish from those that are causally grounded. Consider that a physicist also obtains only singular results from singular experiments conducted in his/her lab, one after the other, but then he/she goes on and subsumes these experiments, and their results under a uniformity and repeatability of natural processes captured by natural laws. The same does not readily appear to be an option for evolutionary biologists, and yet, "The historian is not really interested in the unique, but in what is general in the unique."[64] What does that mean?

Darwin, as others before him, was interested in the question of how a new species originates: what are the causes that lead to the origin of a new species? But the origin of a new species is a historical, indeed a unique event, just as every species itself is unique. A species that originates once and only once in time and space is unique, i.e., unrepeatable, as is the event of its origin. Once a species has gone extinct, it cannot be brought back. It is lost forever, hence the current concern for the loss of biodiversity. But if the origin of a new species is a unique, unrepeatable event, how could it be captured by some natural law, even if that law were of restricted, narrow scope? It seems that the scope of a natural law that would explain the origin of a new species would have to be restricted to such an extent that the law would range over this particular, unique, und unrepeatable event only, and none other. But in so restricting the scope of the law, the explanation would become

[63]Leplin, J. 1997. A Novel Defense of Scientific Realism. Oxford University Press, Oxford, p. 24.
[64]Carr, E.H. 1961. What is History. Vintage Books, New York, p. 80.

uninteresting[65]; it would be reduced to a narrative explanation, a mere description of the origin of a particular species. Such a law would have no generalizing power. Evolutionary biology would be reduced to a historical narrative, i.e., a description of the circumstances in which this or that or any other species originated, or might have originated. Each of these descriptions would be different from any other one to at least some minimal degree, according to the unique historical situation in which any species originated. There would be no access to any general, i.e., common traits of species origination.

However, evolutionary theory can do much better than that. The origin of a new species is, on all accounts, an historical *event*. Once this has become clear, one can proceed to draw a distinction between type events and token events. Token events exemplify the type event of which they are a token.[66] For example, "the Olympic Games in Rio de Janeiro in 2016" is a token event, which exemplifies by a particular instance the type event "Olympic Games." On that basis, one can predict what sorts of disciplines will be part of the Olympic games in Rio de Janeiro in the summer of 2016, and by what rules those disciplines will be played, but one cannot predict the outcome of the games, i.e., who will be the gold medal winners. Who will be the winner of a specific Olympic discipline such as track cycling in Rio de Janeiro in 2016 is a unique event that affords no testable prediction (in Popper's strict sense of the word, although one can of course place bets). It is the rules that govern an Olympic discipline such as track cycling which impart generality on this unique event: the same rules repeatedly govern track cycling in all summer Olympic Games. In a similar sense, the origin of a particular new species is a unique, unrepeatable event that affords no predictability. But there are generalities to speciation events that are captured by explanatory theories, i.e., theories of speciation. Although the origin of a particular new species is a token event, speciation in general – the origin of new species – is a type event, and it is perfectly legitimate to ask the question whether speciation is subject to some general rules. Indeed, an evolutionary biologist may not be so much interested in the unique and unrepeatable origin of a particular species, but rather may study the origin of a particular new species to understand what causal mechanisms underlie speciation events in general. The laws or "rules" that hold generally over speciation events are of a more restricted scope than the laws of planetary motions, but they nevertheless allow the formulation of a small and finite number of models of speciation.

One of those is the allopatric speciation model (allopatric speciation theory), which is perhaps the most popular speciation model amongst modern evolutionary biologists. It is built on the insight that to have new, descendant species originating from an old, ancestral one, the gene flow between the ancestral and descendant populations must somehow be interrupted. For sexually reproducing species, the

---

[65]Kitcher, P. 1993 The Advancement of Science. Science without Legend, Objectivity without Illusions. University Press, Oxford.

[66]Evnine, S. 2001. Donald Davidson. Stanford University Press, Stanford, CA.

gene flow between an ancestral and a potentially descendant population is interrupted if the two populations become separated such that organisms from the
different populations can no longer interbreed. Such separation of populations is
frequently achieved by geographical means. Assume that in the geological past,
vast stretches of a continental land-mass were covered by grassland inhabited by a
particular species of grazing animals. Tectonic activities in the earth's mantle might
result in a subdivision of this originally contiguous grassland area: there could be
the up-folding of a mountain chain, or a marine transgression creating a seaway
across the continental land-mass. The species of grazing animals in question is
anatomically, physiologically, and/or behaviorally incapable of surmounting either
the mountain chain or the seaway. Populations to the east and the west of that
mountain chain, or seaway, would thus become genetically isolated. This provides
an opportunity for the two now separated populations to drift apart genetically:
different random mutations exposed to different selection pressures get fixed in the
two populations. Over time, two new species would have evolved from one
ancestral species as a consequence of tectonic events. The nature of the tectonic
events may be different (mountain building, marine transgression), the outcome of
the speciation event is not predictable (we cannot predict in which way exactly the
two new species will differ from one another, and by which time exactly the
speciation event is complete), but geographic separation is recognized as a general
condition that has to be met for speciation. Consider a continental population of
finches in South America: during a particularly strong storm, a gravid finch female
is accidentally carried to some island archipelago, such as the Galapagos Islands.
There she lays her eggs, and if lucky, the hatchlings will survive and start to build a
new population. This island population is geographically isolated from the continental population, the two populations hence free to genetically drift apart. Over
time, the island population will have evolved into a separate species that can no
longer interbreed with the continental species, and that will have adapted to its new
environment on the island as a consequence of variation and natural selection.
Again, the circumstances are different: it is no longer tectonic effects that separate
populations, but the passive dispersal of a founder of a new population to a remote
area. But geographic isolation is still the general condition that has to be met for
speciation to occur. This is not to say that speciation always and only occurs as a
consequence of geographic isolation: there can be ecological, physiological, or
behavioral factors that can result in reproductive barriers between populations, and
hence trigger speciation. And so we arrive at a "law," or better, a scientific theory of
speciation. Every species is irrevocably unique, every speciation event is likewise
unrepeatable, but (at least some significant degree of) genetic isolation between
populations is a general prerequisite for speciation to happen, and sometimes, but
not always, such genetic isolation is the consequence of geographic isolation. Other
theories of speciation may appeal to ecological, behavioral, or physiological generalities involved in unique speciation events. The result is a hierarchy of laws (in
the appropriately weak sense of the word) of different scope. The requirement of
genetic isolation for speciation to occur is of broader scope than the requirement for
geographic isolation. Geographically isolated populations are genetically isolated,

but not all genetically isolated populations are also geographically isolated: they might be ecologically, behaviorally, or physiologically isolated.

To better understand how causal explanation – in terms of laws or theories of restricted scope – work in science, including historical sciences such as evolutionary biology, it is once more beneficial to return to the notion of universal laws of nature. Let us look at a law as it is used in physics: "For all pieces of magnetized iron, everywhere and always – if an object is a magnetized piece of iron, then it will attract iron filings" under normal physical circumstances. This comes as close to a universal law of nature as seems possible, stating a relation between cause and effect that is said to obtain always and everywhere, throughout the universe, whenever a piece of magnetized iron and iron filings coexist in time and space under the appropriate physical circumstances for the law to be applicable at all. But precisely because it is of such universal nature, the law says nothing about the peculiarities of the "here" and "now." The law says nothing truly, or falsely, about the object that lies on the table before me. In fact, that object lying on the table before me may look so strange that I cannot even say what exactly it is. Or perhaps somebody covered an object on the table before me with a piece of cloth, and asked me to guess what it is. But even if I am completely ignorant of the nature of the object that lies under the blanket, I *can* still say of that object what the law just spelled out, namely that "*if* this were a piece of magnetized iron, *then* it would attract iron filings." I am not saying that the object *is* a piece of magnetized iron; therefore, if it is not, and hence does not attract iron filings, my statement is not false. I also do not need to have myself observed a whole series of regularly recurrent events, such as iron filings being attracted by a piece of magnetized iron, to make that statement. All I need to know is the laws of magnetism, and on this basis I can predict a singular event that could happen here and now, because that event, singular as it is, would be grounded in causal relations if it happened: "*if* the object were a piece of magnetized iron, *then* it would attract iron filings."

This view of looking at causality, or laws and theories, appeals to what philosophers call "counterfactual conditionals." There does not have to be a piece of magnetized iron on the table before me; in fact, there may not be such a piece of metal on the table, which is why my statement can be counter to fact. If the piece of metal lying on the table *is* not magnetized, then my statement *is* counter to fact. But, nevertheless, it there *were* a piece of magnetized iron on the table, *then* it would indeed attract iron filings. If it did not, the laws of magnetism would be in trouble. What this example shows is that laws of nature, or causally grounded theories, have counterfactual force. For laws considered to be universal, such as the laws of magnetism, the counterfactual force is either zero, or one: it is an all-or-nothing affair. A piece of magnetized iron either does, or does not, attract iron filings. Given different degrees of magnetization, the iron filings may be attracted more or less strongly, but attracted they always are – unless the laws of magnetism were wrong. But there are laws of restricted scope, theories that are weaker, such as the theory of allopatric speciation. Unless time is appropriately factored in, geographic isolation itself provides a possibility, but no necessity, for populations to drift apart genetically. It is possible that two geographically isolated populations drift apart to some degree genetically, but merge again at a later time, when the geographic barrier has disappeared (in one of the examples above, a sea-level drop might have removed the trans-continental seaway). As the populations come into contact again, they might still be able to interbreed and merge again. The counterfactual force of the theory of allopatric speciation, therefore, is not an all-or-nothing affair, but comes in degrees, and is time-dependent. The greater the degree of reliability with which a law of nature or theories generate predictions, the greater is their counterfactual force. Some laws are inherently probabilistic, as are the laws of quantum mechanics. Other theories, such as theories on speciation, may make predictions within a range of statistical accuracy only. The counterfactual force of such statistical theories will therefore come in degrees, but as long as better than chance predictions can be generated, the theory will be relevant to scientific discourse.[67]

5.8   Darwin's Test of Evolutionary Theory

Particularly for biological theories, where the exception often confirms the rule and borderline cases are rampant, counterfactual force generally comes in degrees. In most general terms, a counterfactual statement says that "if such and such were the case, then this or that would happen" – but perhaps only with a certain degree of likeliness. Such counterfactual statements can therefore be dissected into a covering law (a theory of a certain scope and possibly of a statistical nature), initial conditions, and a testable prediction. If the laws of magnetism are coupled with the initial condition of the availability of a magnet and iron filings, then the prediction can be made that the magnet will attract iron filings. If the theory of allopatric speciation is coupled with the initial condition of the actual existence of two conspecific populations that have become geographically isolated, then the possibility of speciation obtains, if enough time is allowed for. These are the resources exploited by Darwin, when he argued for the grounding of evolutionary theory in natural causes.[68] The theory of natural selection may not measure up to Newton's Laws in terms of its counterfactual force, but neither is it "the law of the higgledy-piggledy" as Herschel would have it. While not a universal law of nature, evolutionary theory provides resources with considerable counterfactual force. In fact, Darwin himself issued a number of testable predictions in his "*Origin*" (1859): "If it could be shown that our domestic varieties manifested a strong tendency to reversion... I grant we could deduce nothing from domestic varieties in regard to species" (Darwin, 1859, p. 15); "If it could be demonstrated that any complex organ existed, which could not possibly have been formed by numerous, successive, slight modifications, my theory would absolutely break down" (Darwin, 1859, p. 189); "If it could be proved

---

[67]Griffiths, P.E. 1999. Squaring the circle: natural kinds with historical essences, pp. 209–228. In: Wilson, R.A. (Ed.), Species. New Interdisciplinary Essays. MIT Press, Cambridge, MA.

[68]Hull, D. 1999. The use and abuse of Sir Karl Popper. Biology & Philosophy, 14: 481–504.

that any part of a structure of any one species had been formed for the exclusive good of another species, it would annihilate my theory, for such could not have been produced through natural selection (Darwin, 1859, p. 201); "If numerous species, belonging to the same genera or families, have really started into life all at once, the fact would be fatal to the theory of descent with slow modification through natural selection" (Darwin, 1859, p. 302).

Darwin claimed his theory of evolution by variation and natural selection to be causally grounded in the natural course of events. The causes hereby invoked are only the material and the efficient ones, not the formal and final causes that Miller appealed to. The reason is quite simple. Laws or explanatory theories of natural science can be tested and potentially refuted. Laws or explanatory theories of natural sciences fail if the counterfactual condition is rendered actual, yet the consequences specified do not obtain. There may not be a piece of magnetized iron on the table, but may companion can place a certified magnet on the table before me, and if it fails to attract iron filings, then the laws of magnetism are in trouble. Any argument that seeks to establish a relevant role for formal and final causes in natural science must meet the challenge of formulating counterfactual conditionals that would be grounded in such formal and final causes. The challenge has so far not been met. The reason quite simply is that if design, purpose, and goal-directedness should permeate natural processes such as evolution, there is no room left for counterfactual conditionals to fail. Consider the counterfactual conditional a proponent of Creation Science, or Intelligent Design, would have to issue: "If it were the case that. . ., then the doctrine of Creation, or of Intelligent Design would absolutely break down." It seems to be impossible to imagine any state of affairs that could not be explained by the invocation of a Creator, or Intelligent Designer – which is exactly why Laplace claimed that he saw no necessity for such a hypothesis in the context of natural science. Creation, or Intelligent Design, explains everything not only infallibly, but also far too easily. On a creationist account, the world we live in would be the best of all possible worlds, as Leibniz put it – there could be no world possible that would or could be better than ours. Organisms would be perfectly adapted. Alas, they are not. In light of his call for the "best of all possible worlds," the successors of Leibniz had famously to deal with the devastating earthquake that shook Lisbon on November 1, 1755. In light of the Creationist's call for perfect adaptation, Darwin had to deal with variation and competition.

Consider what some authors have called the logical structure of the theory of natural selection, which shows its counterfactual force to be much stronger than that of the "law of the higgledy-piggledy": if it is the case that natural populations tend to grow geometrically (as is revealed by observation), and if it is the case that resources are limited (as is generally the case), then there must be competition for resources within natural populations. It is possible to investigate whether natural populations grow geometrically, and whether in the face of such growth resources are limited: so far, Darwin's beliefs have been sustained by continued research. Given that the research confirms the premises, competition – in some forms and to a greater or lesser degree – *must* occur. Now, if there is competition among organisms, and if organisms are subject to heritable variation, as Darwin had found, then natural selection *must* occur. The consequence is not the best of all possible worlds, but a world of change, of emergence, and innovation. "There is grandeur in this view of life, with its several powers, having been originally breathed into a few forms or into one; and that, whilst this planet has gone cycling on according to the fixed law of gravity, from so simple a beginning endless forms most beautiful and most wonderful have been, and are being, evolved."[69] Using the term only once in his "*Origin*," namely at the very end of the last sentence, Darwin nevertheless radically changed the meaning of the term "evolution."

---

[69]Darwin, 1859, ibid., p. 490.

Chapter 6
The "Law of Superposition"

The 'Law of Superposition' states that in an undisturbed sequence of layers of rocks, the younger layers lie on top of the older layers, such that fossils from deeper layers are geologically older than fossils from some more superficial layers of rocks. The 'Law of Superposition' is subsumed by the Law of Gravity, which is as universal a law of nature was one can wish for. But defending a staunchly empiricist position, Miller denied that the theory of species transformation can be read off from the sequential appearance of fossils in successive layers of sedimentary rocks. His reason was that the Fossil Record does not offer the opportunity to directly observe the process of species transformation. If anything, such a theory of species transformation has to be inferred from the Fossil Record. But for this to be possible, the species – fossil and extant – first have to be classified into a natural system that would reflect their evolutionary relationships. The question then becomes how to classify fossils, or organisms in general, and how to distinguish natural from artificial classifications.

Miller took groups of organisms as they appear in biological classifications as abstract constructs of the ordering human mind, not as historical entities located in time and space. Such classifications cannot reflect the evolutionary history of life on earth, but only the logic of the underlying Plan of Creation – if there ever was one. But even if such an interpretation of biological classification is rejected, it is not easy to identify what is, and what is not, a natural order of organisms that reflects their evolutionary relationships. Horticulturists may classify plants differently from botanists, cooks may classify animals differently from zoo keepers, and ecologists relate organisms to one another in a theoretically relevant way that differs from the relationships researched by systematists and evolutionary biologists. The key here, as Darwin realized, is to distinguish artificial kinds from natural kinds of things that occur in nature.

With Darwin, evolution is not merely a process of species transformation; it also is a process that multiplies species by the splitting of ancestral species lineages. Darwin abandoned the concept of the Great Chain of Being, and replaced it with the branching family tree that translates into a hierarchy of groups within groups. He saw his theory to provide a natural causal explanation for the natural system of plants and animals that had been worked out by the systematists of the Natural History Museums in Paris, London, and elsewhere. The eminent 20th Century evolutionary biologist Theodosius Dobzhansky once said "nothing makes sense in biology except in the light of evolution." The philosophers of biology Kim Sterelny and Paul E. Griffiths recently paraphrased this famous quote as "nothing in biology makes sense except in the context of its place in phylogeny, its context in the great Tree of Life."

O. Rieppel, *Evolutionary Theory and the Creation Controversy,*
DOI 10.1007/978-3-642-14896-5_6, © Springer-Verlag Berlin Heidelberg 2011

6.1   The Superposition of Fossils

Since the mid-eighteenth century, the temporal succession of fossil species was widely recognized. What later became the Primary, Secondary, and the Tertiary Epoch of earth history showed that there had been times when no reptiles lived, and that the reptiles appeared at a time that saw no mammals yet. Even authors who rejected all notions of change and species transformation called the Age of Fishes the "First Creation," the Age of Reptiles the "Second Creation," and the Age of Mammals the "Third Creation." At the Seminari Consiliar in Barcelona, the old collection cases for fossils still carry these designations, or at least did so a few years ago. To say that major groups of organisms came to successively populate the earth in the course of time is neither to pronounce factual statements about species transformation in the course of earth history, nor is it a statement of a universal law of nature. The Law of Superposition merely states that in an undisturbed succession of layers of sedimentary rock, the layer on top is geologically younger than the layer below it. The Law of Superposition is subsumed by the Law of Gravity, and the Law of Gravity is as universal a scientific law as one can wish for. The Law of Superposition is very closely tied to gravity, indeed dependent on it. Descent with modification is a unique historical process and, as we have seen in the previous chapter, quite difficult to capture in terms of natural laws. However, while the formation of sedimentary rocks is a historical process as well, it is a law of nature that rivers flow downhill, not uphill, and that sediments sink to the bottom of the sea, rather than evaporate. The question remains whether the Law of Superposition implicitly allows the inference that in a comparison of "similar" and hence possibly related fossils, those of earlier strata, that is the earlier fossils, must necessarily be considered to represent the ancestors of those found in overlying, that is, geologically younger strata. Are the layers of sedimentary rock equivalent to pages in the book of evolution? Can the process of evolution be directly "red off" from the Fossil Record? Again, the Law of Superposition explains a certain pattern of order, the succession of fossils through time as the consequence of sedimentation; it does not explain that pattern of order as a consequence of descent, with modification.

In 1846, the American Journal of Science published a review of Chambers' book *"Vestiges of the Natural History of Creation"*: "Geology, if its facts mean anything, fully shows that tribes of animals have successively disappeared, owing to physical causes; and that the new races have appeared by creation, and not by gradation, or 'progress'."[1] "Gradation" means the stepwise transformation of species; "progressive development" means progress in the series of species transformation along the Great Chain of Being: always adding to and building upon what has been achieved by previous steps of transformation, species transformation would trend toward ever higher levels of perfection, as was reflected in the ladder of life, running from mushrooms to humans. If this was to be the result of an historical process fully

---

[1]Anonymous, 1846. Sequel to the Vestiges of Creation. American Journal of Science, Ser. 2, 1: 252.

determined by natural laws (such as Chambers' Law of Development), where an unbroken chain of causes and effects prevailed, then an absolutely uninterrupted sequence of graduated steps of transformation would have to be predicted and demonstrated. Darwin was not the first to point out that gaps in the Fossil Record might be used by his opponents as an argument against species transformation: "He who rejects [the imperfection of the geological record] will rightly reject my whole theory."[2] The significance of gaps in the succession of fossils had been recognized before by Hugh Miller, who used it in his criticism of Chambers' book. If gaps separate cause from effect, the determination of the natural course of events is incomplete, and a window opens for chance to become part of the game – or, conversely, for the claim that secondary causes can be temporarily suspended to allow for a special intervention of the First Cause. Chambers' answer was, of course, that this is an anthropocentric view of the Creator, where an eternal entity is supposed to step into time and space.

It is generally acknowledged that the natural circumstances that even only potentially allow the fossilization of plants and animals are rare and far between, that the actual fossilization of a plant or animal, therefore, is a rather exceptional event, and that the actual finding of fossils is even more exceptional. Gaps in the Fossil Record are thus to be expected, further enhanced by other factors such as erosion, which through eons of time destroyed sedimentary rocks and their fossil content as a consequence of gravity. But, as we have seen, Miller insisted on a direct perceptional access to the records of earth history, and leafing through the layers of sedimentary rocks he found gaps in the succession of fossils that seemed to refute any theory of species transformation. Ironically, Miller, who wanted to base all science on brute observation, appealed to perceptional access to something that is not there, to something that is missing, to negative evidence that is, the missing intermediate fossil, in his refutation of Chambers' vision.

In contrast, Chambers was impressed by the succession of fossils through time. Simple "types of construction," as the German Natural Philosophers used to say, appeared in lower and hence earlier layers of sedimentary rock than more complex types of organisms. The temporal succession in the appearance of the classes of vertebrate animals seemed to provide a perfect example supporting this generalization. Fossil fishes precede tetrapods, reptiles precede mammals, tetrapod mammals precede bipedal humans in the succession of sedimentary deposits. Chambers felt that this succession in time of ever more complex types of organization would reflect the doctrine of progressive development, a series of species transformations which unfolds through time and space on the basis of secondary causes enacted by the First Cause. Miller accepted Chambers' overall assessment of the vertebrate Fossil Record as documenting a progression toward what were considered to be ever higher levels of organization, but going beyond the argument of the incompleteness of the Fossil Record, he pointed out that the most advanced types of one class were not necessarily and immediately followed in the succession of deposits

---

[2]Darwin, Ch. 1859. On the Origin of Species. John Murray, London, p. 342.

by the lowest representatives of the next class to follow on the ladder of life. Although he agreed that there existed an overall trend to higher organization within vertebrate animals, he maintained that their successional appearance in the Fossil Record, if analyzed with enough concern for detail, violates the assumption of a gradual process of change. Scrutinizing the Fossil Record, he cited examples believed to document the fact that within one class such as fishes, for example, more complex types of organization such as his *Asterolepis* could appear before the appearance of less complex types of organization such as sharks. This argument requires, of course, some measure of complexity, and such a measure is notoriously hard to come by. *Asterolepis* is, indeed, a bizarre looking creature all encased in a heavy bony armor. Sharks lack such formidable protection, but they also do not need it, in fact cannot afford it, as it would make them too heavy and clumsy for fast movement through a dense medium such as water. *Asterolepis* was a bottom-dweller, whereas sharks are free-swimming predators. Even so, and allowing for some charity in the interpretation of his writing, Miller may have been partially right in his observations based on the evidence available to him, but he was wrong in his conclusions. If derived members of a group are observed to appear before less derived representatives of that same group, this might still be attributed to the incompleteness of the Fossil Record. Perhaps, archaic fishes lived before *Asterolepis*, but had just not yet been found – as was indeed to be revealed by later research. Even worse, before one could judge the doctrine of progressive development on the basis of the Fossil Record, one had to classify the rocks in which those fossils occur. Although the Law of Superposition is very straight forward in its explanation of the relative age of depositional strata, the classification of these successive layers of rock proved far less easy in practice, and was, indeed, still debated at the time of Miller's writing.[3] Miller, therefore, looked for more convincing arguments, and found them in the field of biology known as systematics. Systematists are concerned about which organisms form a natural group and which do not, and also how these natural groups, once discovered, are related to each other. To say that mammals are more "complex" in their anatomy and physiology, or more "highly evolved," or more "derived" than reptiles makes only sense if mammals on the one hand, reptiles on the other, from a natural group. For that statement to make sense, it is further required that mammals and reptiles are somehow related to each other, be that in the Creator's mind or through evolution. Miller, therefore, set out to investigate the question as to how, in fact, we know which animals form which group, and how groups are related?

---

[3]Secord, 2000, Victorian Sensation. The Extraordinary Publication, Reception, and Secret of *Vestiges of the Natural History of Creation*, The University of Chicago Press, Chicago, p. 243f; see also Rudwick, M.J.S. 1985. The Great Devonian Controversy: The Shaping of Scientific Knowledge Amongst Gentlemanly Specialists. The University of Chicago Press, Chicago; Rudwick, M.J.S., 2008. Worlds Before Adam. The Reconstruction of Geohistory in the Age of Reform. The University of Chicago Press, Chicago.

6.2   Systematics and the Classification of Organisms

What, Miller asked, was progressive development after all, and how could it be measured? Indeed, if vertebrate animals are classified such that reptiles follow fishes, mammals follow reptiles, and humans follows tetrapod mammals in turn, then this classification would mirror the arrow of time pointing in the direction of progressive development. But how do we measure complexity, how do we measure progress: is a lobster more, or less, complex than a shark? In Miller's view, such considerations only raised the question what principles would form the starting point for animal classification, and how these principles could be defended? How do we know that reptiles are more primitive than mammals (if indeed they are), and that mammals are related to and, in fact, descended from reptile-like ancestors? How do we know what a fish is, and why should fishes not be ancestral to birds, as, indeed, is claimed in the Book of Genesis, with flying fishes providing the missing link? If transformation was as gradual and continuous as proposed by Chambers, where was the line to be drawn between reptiles and birds, between mammals and humans? What are "groups," such as fishes, or birds, linked as they would have to be by an unbroken series of intermediates, and how could we recognize them?

There is a seemingly easy answer to this question, in that it is the characters that make groups. Fishes have scales, birds have feathers, and mammals are characterized by the possession of hair. True enough, this is a possible way to diagnose groups – but there are alternatives, just as there are other characters. Chambers, for example, had not attributed much importance to scales, feathers, and hair, but had mentioned the importance of the color of body fluids, particularly of blood, in the assessment of the relationships between animals with and without a backbone. Miller, on the contrary, used another one of Chambers' preferred characters, namely relative brain size, in his refutation of continuity in progressive evolution. Today, comparative biology strives for a classification that would be the most "natural" one, revealing the "hidden bond of community of descent"[4] as Darwin put it. Evolution, however, is not the only and exclusive perspective from which things can be classified. Indigenous people, for example, may classify the contents of their natural environment from an entirely different point of view than modern biological science does.[5] They might partition the content of the living world into what is edible and what is not, as cooks also do. Or their partitions might concern healing powers, or the lack thereof, as do the partitions of pharmacists. From such

---

[4]Darwin, 1859, ibid., p. 426.

[5]Berlin, D., D.E. Breedlove, and P. H. Raven. 1966. Folk taxonomies and biological classification. Science, 154: 273–274. Brown, C.H. 1985. Modes of subsistence and folk biological taxonomy. Current Anthropology, 26: 43–64. Atran, S. 1999. The universal primacy of generic species in folkbiological taxonomy: Implications for human biological, cultural and scientific evolution, pp. 231–261. In: Wilson, R.A. (Ed.), Species. New Interdisciplinary Essays. The MIT Press, Cambridge, MA.

observations, the question necessarily arises as to who got it right or wrong? Is it Chambers, Miller, indigenous people living on the banks of Sarawak River, the cook, the pharmacist, or the biologist working in a modern Natural History Museum? Chambers certainly pursued an agenda, but so did Miller, just an opposing one. In different contexts of investigation and argumentation, different characters are deemed important, different groupings are obtained, different conclusions drawn.[6] Indigenous people might classify the contents of their world within the context of economic or medical interests, following their intuition and the experience of the elders. Modern museum taxonomists classify organisms on the basis of vast databases based on DNA sequences, which renders computer support in the analysis of evolutionary relationships a necessity. That may sound like a modern solution to an old problem. In reality, however, it only engenders debates about how to align strings of DNA, and which algorithms and computer programs to use in the reconstruction of the Tree of Life.

The simple fact of a succession of fossils through time cannot provide a direct (i.e., observational) clue to the history of life on earth. For the Fossil Record to make sense, it must first be ordered according to the theories and methods of systematic biology; the fossils must be correctly identified, related to one another as well as to groups of extant organisms and classified accordingly, before their succession through time starts to make sense in evolutionary terms. As argued by Darwin, evolution is a theory that provides a causal explanation for the order that prevails amongst organisms. Or, to put it in other words, descent with modification is the causal explanation Darwin offered for the graded similarity and order that was worked out by the systematists of the Natural History Museums in Paris, London, and elsewhere. Chambers supported his doctrine of progressive evolution by what he observed and believed to be the true order of nature; however, Miller saw things differently and hence reached different conclusions. What then is a natural group, indeed a system of natural groups calling out for an explanation in terms of the origin of new species, in terms of descent with modification? Or, to put it the other way around: if it is the case that systematists find plants and animals to form what appears to be a hierarchical system of natural groups, and when paleontologists find that the Fossil Record lends a congruent time dimension to such a system, how then are we going to explain this system? Darwin thought that evolutionary theory is the natural answer to that question. However, this answer is valid only if the hierarchical system that is reflected in biological classifications is a natural one, and not an artificial one such as the one based on economic interests of a human society – but how can we know the difference?

Miller insisted: "Further be it remarked, that the scheme of classification which gives an abstract standing to the Chondropterygii, is in itself merely a certain perception of resemblance which existed in certain minds, having cartilage for its

[6]Dupré, J. 1993. The Disorder of Things. Metaphysical Foundations of the Disunity of Science. Harvard University Press, Cambridge, MA.

general idea..."[7] This statement looks rather innocent on a first reading, but through it Miller defended a certain philosophical approach to animal classification that is absolutely incompatible with a theory of species transformation. Recall Louis Agassiz' mentorship for Miller, and consider what, a few years later, Agassiz had to say about the zoological system in his *"Essay on Classification"*: "What we call a branch expresses, in fact, a purely ideal connection between animals, the intellectual conception which unites them in the creative thought"[8] (i.e., in the thought of the Creator). Agassiz sketched a peculiar, indeed philosophically challenging, vision of the natural system: he thought that natural groups exist in nature, for us to discover, yet the reality of these groups he rooted in the thought of the Creator, not in theories of species transformation.[9] In contrast, and for Miller, the "Chondropterygii" had a merely "abstract standing," which seems more in tune with the idea that they represent a "general idea" grounded in the blueprint of Creation. Let us, therefore, unpack Miller's statement carefully, as it is of great importance for an understanding of whether or not, and under which circumstances, the work of systematists can be related to a theory of evolutionary change.

6.3 Ideal vs. Natural Systems

First, Miller invoked a group of organisms (i.e., the Chondropterygii or chondrichthyans as they are called today) to which belong the sharks, skates, and rays, and their fossil relatives. But to Miller, this is not a natural group, not a historical entity marked out by common ancestry, that is, one with a definitive beginning in time and a locus of origin, but instead one that has merely "abstract standing" in "certain minds." True to his empirical approach to nature, Miller invokes observation as the basis of all scientific discovery, but then claims that it yields nothing more than a "perception of resemblance." On that account, chondrichthyans do not share a common history by virtue of their common ancestry, but only some resemblance, a similarity furthermore that exists not in nature due to causal mechanisms of inheritance, but in "certain minds" instead. For Darwin, as well as for modern systematists, a natural group such as the one formed by sharks, skates, and rays, if indeed it is natural, comes into being through descent with modification at a certain time in earth history and in a certain place. Such a group is introduced into time and space through its origin from a unique common ancestor, just as such a group, or its parts, can exit from time and space through extinction. Ideas cannot

---

[7]Miller, H. 1849 [1850], Foot-Prints of the Creator: or, the *Asterolepis* of Stromness, Agassiz, L. (Ed.). Gould, Kendall and Lincoln, Boston, p. 151.

[8]Agassiz, L. 1859. An Essay on Classification, Longman, Brown, Green, Longmans & Roberts, London, p. 218.

[9]Winsor, P.M. 1991. Reading the Shape of Nature. Comparative Zoology at the Agassiz Museum. The University of Chicago Press, Chicago. See also Rieppel, O. 1988. Louis Agassiz (1807–1873) and the reality of natural groups. Biology & Philosophy 3: 29–47.

come into being in the same way a new species can through a process of speciation. Ideas also cannot go extinct in the same way a species can. An extinct species is lost forever, but a forgotten idea can be rediscovered. No lesser minds than those of the philosophers Gottlob Frege (1848–1925) and Karl Popper have argued – controversially of course – that although it is possible that nobody thinks of a particular idea at a particular point in time and space, this does not mean that the idea does not exist in some different realm, one unconstrained by time and space. The same would hold for theories: Popper insisted that Einstein did not invent the (special and general) theory of relativity – he *discovered* it, but for it to be discovered, it had to have an objective existence. Thoughts for Frege, laws of nature for Popper, existed in a "third realm," ready to be grasped by the inquiring mind.

Chondrichthyans such as sharks, skates, and rays are also known as cartilaginous fishes, because their internal skeleton does not ossify: they do not have bony vertebrae and ribs; instead, the vertebrae and ribs of modern sharks consist of calcified cartilage. It is this similarity, the fact that their internal skeleton remains cartilaginous in the adult, which according to Miller marks out the group Chondrichthyes, yet not in terms of a trait inherited from a common ancestor, but instead, and according to Miller, in terms of a "general idea." On this account, it is not a historical process of evolution that resulted in a group of fishes retaining a cartilaginous internal skeleton in the adult. Instead, it is the observing and comparing naturalists who in their minds develop the concept of a group of fishes that all retain a cartilaginous internal skeleton in the adult. Such sharing of an internal cartilaginous skeleton in the adult organism is a "general *idea*," according to Miller, one that is exemplified by sharks, skates, and rays; it is not a *fact* of natural history rooted in genealogical relationships. Accordingly, all sharks, skates, and rays form a *class* of living organisms, and this class is defined by the universally shared property of having a cartilaginous internal skeleton as adults. Chondrichthyans thus exemplify a *type* of organization, yet that type is not located in nature (i.e., in space and time) but in the minds of biologists instead (i.e., in the logic of their classifications). For Darwin and his successors, it is their common evolutionary history that ties sharks, skates, and rays together as a natural group. The fact that all organisms that form part of this group lack an ossified internal skeleton at their adult stage is a trait that the group inherited from its most recent common ancestor. For Miller, a group such as chondrichthyans is a "mental concept": the group is the result of a taxonomist's mind ordering nature, thus tracing the footsteps of the Creator. The taxonomist lets his mind roam through nature, and in so doing the mind of the taxonomist rakes into the class called Chondrichthyes all the organisms that are found to lack a bony internal skeleton in the adult. The lack of a bony internal skeleton in the adult thus is not a property believed to be inherited from the chondrichthyan ancestor, but one that the systematist uses to *define* the class of cartilaginous fishes called Chondrichthyes. We cannot define the world, we can at best define the meaning of words. Used as a *defining* character, the lack of a bony internal skeleton in the adult cannot mark out a natural group of common ancestry. It can only define the meaning of the name "Chondrichthyes," that is, the concept of cartilaginous fishes that the systematist in his mind associates with that name. The taxonomist defines the concept of

Chondrichthyes in her mind, and then looks out into the world trying to see what does, and what does not, fit the concept based, as it is, on defining properties. Such defining properties are also known as essential properties, which mark out classes that cannot also be historical entities, as natural groups would have to be. Essential properties cannot change; they are timeless, universal. On the basis of the logic of class membership, the latter being defined by a shared essential property, the class of cartilaginous fishes cannot change, for such change would require its defining essential property to change; yet the property cannot change, because it defines the meaning of the name Chondrichthyes. The essential property defines the conditions of membership in the class of cartilaginous fishes. If there is a fish whose internal skeleton is cartilaginous at a juvenile stage, but bony (ossified) in the adult, then that fish by definition cannot be a chondrichthyan. On Miller's account, cartilaginous fishes turn out not to form a natural group that shares a common historical (evolutionary) ancestry, but to be an abstract class instead that lies beyond time and space, satisfying not a theory of species origination and transformation, but the timeless logic of the ordering mind of the systematic biologist instead.

Miller viewed the classes of the animal kingdom that are recognized by biologists as conceptual abstractions rather than as historical entities located in nature. According to Miller, the groups and subgroups of the "natural" system constituted logical constructs of the ordering mind, and not historical groups produced by a natural process of species origination and transformation. If cartilaginous fishes (chondrichthyans) are a natural group rooted in the process of evolution, there is the possibility for them to go extinct. Should that happen, they would be gone, irreversibly lost forever. They would share the same fate as did the mammoth on George Cuvier's account. Cuvier, remember, denied the "degeneration" of the mammoth into the modern elephant, but called the mammoth instead "*une éspèce perdue*," a lost species. Such a species cannot be brought back once it is lost through extinction. If, as Miller believed, chondrichthyans are a logical class, then this class is timeless: its members can go extinct, but the class cannot. By analogy, there may be no mammoth alive anymore today, but the *class* of all mammoths that ever existed on earth still exists today – not in nature, of course, but in the mind of paleontologists that classify fossils along the lines of Miller's arguments. Take Chondrichthyes to be a natural group, rooted in a unique common evolutionary origin, and further imagine (counterfactually) all chondrichthyans that lived 100 million years ago to have gone extinct at that time. The natural group of chondrichthyans would have gone extinct, would have become a "lost group," a 100 million years ago. That group would have been irretrievably lost! Imagine that through some exceedingly improbable series of evolutionary events, a group of fishes evolved from a unique common ancestor 100 years ago that shares all the characteristics of the extinct chondrichthyan fishes, including the absence of a bony internal skeleton in the adult. This newly evolved group of cartilaginous fishes could not be classified with the extinct chondrichthyans in a natural system, because the two groups of cartilaginous fishes would not share a common evolutionary origin. This problem does not obtain for a created world.

If "cartilaginous fishes" form a logical class, all the members of which had gone extinct 100 million years ago, there would be no problem to assign to this same class cartilaginous fishes that were newly created 100 years ago, for they would still exemplify the same idea of the Creator. Imagine that the chondrichthyans that went extinct 100 million years ago were created by God on the basis of a blueprint, a mental concept that includes the absence of a bony internal skeleton in the adult as a defining characteristic. The Creator had it arranged so that all the members in this class of cartilaginous fishes would have disappeared 100 million years ago. But then, 99,999,900 years later, the Creator stepped into time and space again and – using the same old blueprint – created jawed fishes again that retain a cartilaginous skeleton in the adult. These new creatures would without any difficulties fit into Miller's class "Chondrichthyes," since they would share its defining property. On Miller's account, the temporal succession of the appearance of classes in the Fossil Record cannot be explained as a result of a natural process of evolution, but is, instead, incompatible with the idea of a causal – historical process of descent with modification. Causal relations require the existence of objects in time and space: only existing things (objects) can take part in causal relations. But classes exist in the realm of logic, if at all, not in nature. Think of the class of prime numbers, for example: its members (i.e., any particular prime number) cannot take part in a causal-historical process. Biologists talk about the danger of loosing a species through extinction as a consequence of habitat destruction. Mathematicians have hardly any reason to be concerned about the possibility of extinction of the class of prime numbers, or of any of its members. You can go outdoors and kick rocks around; try to kick any member of the class of all prime numbers around! The Fossil Record exemplified, for Miller, a certain logic of order, of pattern, that can be captured in a logically structured classification of classes within classes, all of which are timeless mental concepts defined by essential properties and ultimately rooted in the mind of the Creator. Even more, all that was demonstrated by the parallelism of order in the classification of organisms, in their embryological development, and in their temporal appearance in the Fossil Record, was the equivalence of the underlying ordering principle, which, through its perfect harmony that is evident in classification, embryology, and earth history leads back to the rationality of the First Cause, the blueprint of Creation conceived in the mind of God. Or so Miller thought. The "three-fold parallelism" of classification, embryology, and paleontology revealed, in Miller's view, design, purpose, and goal-directedness in nature: "...and the arrangement seems at once a very wonderful and a very beautiful one. Of that great and imposing procession of being of which this world has been the scene, the program has been admirably marshaled. But the order of arrangement in no degree justifies the inference based upon it by the Lamarckian."[10] Lamarck, recall, is generally acknowledged to be the first author to have proposed, and defended, a complete theory of species transformation early in the nineteenth century.

---

[10]Miller, 1850, ibid., p. 228.

Darwin, in contrast, correctly recognized imperfections in the three-fold parallelism. Even more importantly, he recognized serious problems with the Great Chain of Being, which is a linear arrangement of plants and animals along a gradient of increasing complexity. It is not a branching system, not a family tree. Species could conceivably transform and through such transformation climb up the ladder of life, as Chambers had argued. But such species transformation along the Great Chain of Being could not explain the multiplication of species. For Darwin, evolution meant not merely the transformation of a species into another, supposedly more highly evolved one. For Darwin, evolution made two or more out of one, as indeed the number of species has increased through time. Species are lineages of ancestral and descendant populations, and if such a lineage splits, two species evolve from one. Furthermore, some of the most primordial forms of life still populate the earth, such as blue algae. How could that be, asked Darwin? Why had they not climbed up the ladder of life through transformation? But if they had not, yet other organisms reached higher levels of complexity, then there must have occurred a multiplication of species, just as there is a multiplication of family members over time. The history of a family is represented as a branching tree, not as a ladder. In one of his early notebooks dating back to 1837, Darwin sketched his first crude branching diagram, where lines split and split again. It is not the picture of classes on top of classes stacked up to form a progressive ladder of life that matches evolutionary progress. It is, instead, the picture of evolving lineages that split and split again that best represents the evolutionary process.

6.4   Artificial vs. Natural Kinds

With this insight into evolution, Darwin added an entirely different twist to this argument: "Why out of the thousands of forms should they all be classified. Propagation *explains* this."[11] He went on to point out: "My theory *explains* that *family* likeness, which as in absolute human family is indescribable yet holds *good*, so does it in real classification."[12] What Darwin is saying here is that if we restrict ourselves to merely "perceived similarity," which is then used to build mental constructs and nothing else, there are innumerable ways to classify organisms. One of them would be the Great Chain of Being, another one would be the family tree. Cooks may classify animals in different ways than zoo keepers, horticulturists may classify plants in different ways than botanists, and ecologists may classify organisms in different ways than systematists. Why, then should biologists have settled on a single scheme of classification, or at least strive to converge on a single

---

[11]DeBeer, G. 1960. Darwin's notebooks on transmutation of species, part I. Bulletin of the British Musuem (Natural History), Historical, 2: 55. Emphasis added.

[12]DeBeer, G., 1960. Darwin's notebooks on transmutation of species, part II. Bulletin of the British Musuem (Natural History), Historical, 2: 97.

hierarchical arrangement that captures a branching Tree of Life? There must be reasons for this, and these better entail an appeal to natural causes, for without recourse to natural causes, there is no basis on which to judge any classification as being a natural one. If all we have to go by are perceptions of certain relations of similarities that then give rise to abstract groupings in the observer's mind, there is no basis on which to chose a "natural" one from the multiplicity of schemes of grouping that are possible. Mere "raw" similarity is often claimed to lie in the eye of the beholder. Philosophers have long commented on the elusive nature of "similarity": Nelson Goodman called it "a pretender, an imposter, a quack."[13] Kim Sterelny and Paul Griffiths pointed to pigs and oysters that share the property of not being considered kosher food by orthodox Jews; that property is a similarity shared by pigs and oysters – should oysters therefore be classified with pigs?[14] Should the naturalist be interested in deciphering the family tree of birds, and to classify birds accordingly, or should be content to follow the cook who puts all "waterfowl" into one basket, based on shared gustatory properties? Should the botanist try to decipher the family tree of plants, or should she be satisfied following the ecologist who groups vastly different sorts of plants in a category called "tropical rain forest," based on perceptual geographical and climatic conditions? Systematic biology puts lions and tigers into the group of cats; foxes and wolves in a group of dogs. Why not just put them together with all other meat-eaters into a group of carnivores? Pythons, eagles, wolves, and tigers are all carnivores, sharing perceptual properties such as the behavioral property of killing prey to make a living. That is a perfectly good reason to put them together in a group, called the carnivores. Yet most people, and certainly biologists, feel more comfortable to group the snakes with other reptiles, the eagle with other birds, wolves with other dogs, and tigers with other cats. This is also how zoos are organized. Zookeepers and the visiting public seem to agree that exhibiting the snakes in the reptile house, the eagle in the bird house, and the wolves and tigers with other mammals seems more natural than if they were all exhibited in a "house of carnivores." But what kind of naturalness is relevant in this context? Darwin's answer was that groups have to be natural in the sense of being rooted in causal natural processes, and the causal process most important to him was descent with modification. There is absolutely nothing wrong with ecologists making use in their ecological theories of concepts such as "carnivores," "tropical rain forests," or "coral reefs" – for such causally grounded concepts can be relevant to ecological theory construction. After all, people talk about the ongoing deforestation of the Amazon basin, or the destruction of the Great Barrier Reef. But for the phylogenetic systematists who seek to reconstruct the Tree of Life, the web-like relations researched by ecologists are not the most important ones. In phylogentic systematics, the most important relation is the one of descent with

---

[13]Goodman, N. 1972. Seven strictures on similarity. In: Goodman, N. (Ed.), Problems and Projects. The Bobbs-Merrill Company, Indianapolis, p. 437.

[14]The example is from Sterelny, K., and P. Griffiths. 1999. Sex and Death. An Introduction to Philosophy of Biology. Chicago University Press, Chicago.

modification, and these relations are best pictured not by a ladder but by a branching tree (at least in the case of most multicellular organisms; some organisms such as bacteria show more web-like evolutionary relations that result from horizontal gene transfer[15]). The similarities that group plants and animals into natural groups must be perceptual, but cannot be only perceptual: they also need to be grounded in the causal relations that govern the course of evolution. Natural groups so grounded are not merely abstract mental constructs, however, as Miller would have it, but natural in the sense of being historical, that is, located in time and space. The abstract groups that Miller defended are constructs of the ordering human mind. The natural groups that Darwin appealed to engage in causal natural processes, or result from prior such engagement, just as the token tigers engage in the processes of predation, competition, reproduction, etc.

In order to keep things simple, let us concentrate on sexually reproducing organisms. The most important causal evolutionary process these organisms engage in is reproduction. It is through reproduction that variation is introduced into a population, and it is through reproduction that many are produced from two. Therefore, thanks to reproduction, the theory of natural selection gains purchase in the dynamics of natural populations, and it is natural populations that provide the cradle for new species to evolve. Individual tigers engage not only in competition for reproductive partners, and in reproduction itself, but also in predation, killing prey to make a living, and feed their offspring. To engage in predation is to engage in a causal process, and there again natural selection will reward success: more food for more offspring carrying the successful genes (i.e., on average). For sexually reproducing organisms, the interbreeding population is the most basic unit in the hierarchy of natural groups. Now, tigers do not interbreed with pythons and eagles, but like tigers, pythons and eagles engage in the causal process of predation. What is then the difference between the species of tigers, *Panthera tigris*, composed as it were of interbreeding populations, and the group called carnivores, composed of pythons, eagles, tigers, and wolves? The species *Panthera tigris* finds its natural grounding in the causal process of interbreeding, ultimately in the causal process of its evolutionary origin; the group of carnivores finds its natural grounding in the causal process of predation. Should the species *Panthera tigris* be considered a real historical entity in nature, the group of carnivores only a mental abstract, a theoretical construct? What is the difference between a species such as *Panthera tigris* and an ecological grouping such as carnivores? The difference is that "carnivores" exemplify an ecological kind that is relevant to ecological theory construction. In contrast, the species of tigers (*Panthera tigris*) is a historical entity (a causally integrated system), located in space and time, that also exemplifies a

[15]Doolittle, W.F. 1999. Phylogenetic classification and the universal tree. Science, 284: 2124–2128. Doolittle, W.F. 2009. The practice of classification and the theory of evolution, and what the demise of Charles Darwin's tree of life hypothesis means for both of them. Philosophiocal Transactions of the Royal Society of London, B 364: 2221–2228. Doolittle, W.F., and E. Bapteste, 2007. Pattern pluralism and the Tree of Life hypothesis. Proceedings of the National Academy of Sciences, 104: 2043–2049.

historically conditioned genealogical kind that is relevant to evolutionary theory construction.[16]

The eminent twentieth century philosopher Willard Van Orman Quine stated that it comes natural to us to sort similar things into kinds.[17] The philosopher Thomas S. Kuhn noted that this is after all how all the children subdivide the world into "dogs and cats, tables and chairs, mothers and fathers."[18] The kinds that children subdivide the world into were called "innate nominal kinds" by the philosopher John Dupré[19]: "innate," because it comes natural to the children to subdivide the world into tables and chairs, cats and dogs, "nominal" because these kinds, or concepts, are abstract, just as Miller thought the kind, or concept, of cartilaginous fishes to be. Although such classifications take children a long way to cognize and organize their world, they are not good enough for science. Science seeks to fulfill Plato's requirement to "carve nature at its *real* joints", i.e., to discover the *natural* kinds that are grounded in natural causal processes, which are also the sort of kinds that can be and are relevant to scientific theory construction. Natural kinds therefore are kinds of things, or kinds of stuff, that occur in nature. Tigers and elm trees are two different kinds of things that occur in nature, water and gold are two different kinds of stuff that occur in nature. Once chemists have explained to us the molecular structure and the consequent causal dispositions of "water," we understand why water freezes at zero degrees Celsius, and we can predict that it would do so in another possible world like a future one. Having been educated about the causal properties of the H_2O molecule, I can confidently state that "if this pot that stands on the table over there were to contain water, then that water would freeze if it were cooled down to zero degrees Celsius." Once geneticists have explained to us the genetic makeup and the consequent causal dispositions of tigers, we will understand that tigers will give birth to other tigers or something close, but not to polar bears, and we can predict that this will be so.

Take the childrens' distinction of tables and chairs, cats and dogs. It is intuitively evident that there is a difference between things such as tables and chairs, and other things such as cats and dogs. Cats, such as tigers, and dogs, such as wolves, naturally occur in nature. But whereas a picnic table may occur in nature, in a forest preserve, for example, it does not occur there naturally. The table is built according to plan and purpose, as is the bench next to it. At one point, there must have been a carpenter, who drew a blueprint of what he considered to be the most

---

[16]See discussion and references in Rieppel, O. 2007. Species: kinds of individuals or individuals of some kind. Cladistics, 23: 373–384. Rieppel, O. 2009. Species as a process. Acta Biotheoretica, 57: 33–49. Rieppel, O. 2009. Reydon on species, individuals and kinds: a reply. Cladistics, DOI: 10.1111/j.1096-0031.2009.00290.x

[17]Quine, W.V.O. 1994. Natural kinds, pp. 42–56, In: Stalker, D. (Ed.), Grue. The New Riddle of Induction. Open Court, La Salle, IL.

[18]Kuhn, T.S. 1970. Logic of discovery or psychology of research, pp. 1–23. In: Lakatos, I., and A. Musgrave (Eds.), Criticism and the Growth of Knowledge. Cambridge University Press, Cambridge, UK; reference is to p. 17.

[19]Dupré, 1993, ibid., p. 268.

practical picnic table and then proceeded to execute that plan to reach his goal, which was to put a picnic table in the forest preserve for people to use. Tables are sort of a kind, as are chairs, but they are artificial kinds. They are man-made kinds, their tokens – the token table or the token chair – built according to design and for a certain purpose. This is not the case for natural kinds. The tokens of natural kinds – the sample of water or gold, the token tiger or elm tree – take part in natural causal processes, which are not known to operate according to design and purpose. There is nothing in the fundamental laws of physics that indicates plan and purpose. What the fundamental laws of physics tell us, at least on some interpretation, is that the laws of nature are, in an important way, statistical. The fundamental laws of physics tell us that a Laplacean demon is impossible. Even if there were such an omniscient being who knew all the relevant laws of physics and all the relevant conditions all atoms of this world are presently in, that being would still not be able to predict even the immediate future of this world in every detail. Such is the insight from the fundamental laws of physics, and the same is true for the fundamental theories of biology. There is nothing in Darwin's or current versions of evolutionary theory that supports the idea of perfect adaptation according to design and purpose. The natural kinds that are relevant to evolutionary theory construction are not marked out by unchanging essential properties in virtue of which they are governed by universal laws of nature. Instead, the natural kinds that are relevant to evolutionary theory construction are marked out by some sort of family resemblance that is the result of fundamentally statistical laws of biology, such as those formulated by theories of inheritance, as are theories of population genetics, for example.

But how, then, do biologists "carve nature at its *real* joints"? How do they discover the "real" similarities in nature, those that are grounded in causal processes and hence mark out natural (as opposed to artificial) kinds that provide the basis for the reliable inference of biological theories such as the theory of evolution? With his comments on the group of cartilaginous fishes, Miller hit a major point of contention. If it is true that anything can be classified in a variety of ways to conform to a multitude of purposes, then each such classification is a conceptual construct, and none can claim to be the only one reflecting the past history of descent with modification. Most notorious in that regard was William Sharp MacLeay's "quinarism," a theory put forward in 1821 that sought to group all organisms in circles of five. MacLeay was a British amateur entomologist[20] who, after his emigration to Australia, presented his ideas on classification in two volumes: the first, a discussion of scarab beetles (1819); the second, an elaboration of his method of classification (1821).[21] According to his theory, all animals could be arranged, on the basis of their affinities, in circles, each composed of five taxonomic units. "Osculating" taxonomic units would connect adjacent circles,

---

[20]Panchen, A.L., 1992. Classification, Evolution, and the Nature of Biology. Cambridge University Press, Cambridge, UK. Winsor, M.P. 1976. Starfish, Jellyfish, and the Order of Life. Yale University Press, New haven, p. 82.

[21]MacLeay, W.S., 1819, 1821. Horae Entomologicae: Or, Essays on the Annulose Animals, 1 vol. in two parts. S. Bagster, London.

thus bridging the gap between adjacent circles as was required by the principle of plenitude,[22] and the continuity of forms it implies. One of his examples of such an "osculating" group was the barnacles (Cirripedia), which would connect the "Radiata" (echinoderms) with the Annulosa (arthropods). MacLeay's quinarism enjoyed great popularity in England from the 1820s to the 1840s[23] and was consequently accepted in the first edition of "*Vestiges*" by Chambers. The possibility of such a strict classification of the animal kingdom into interconnected circles of five was considered proof of the fact that nature was ordered according to some universal law that would reflect back on the First Cause.[24] James A. Secord recounts the story of how after perusing the first edition of "*Vestiges*" in the British Museum library, Darwin realized that he would have to deal with the issue of classification after all[25], for MacLeay's "quinarism" certainly did not seem compatible with evolution through natural selection.[26] It seems, indeed, reasonable to suggest that once Darwin had realized the necessity to refute the Quinary System[27], he plunged himself into the systematic study of barnacles to expose the artificial nature of MacLeay's method of classification.[28] Although Louis Agassiz credited MacLeay with the insight of distinguishing between affinity (homology in modern terminology) and analogy (independently acquired similarity)[29], the question still remains of how to get from an artificial classification as was his "quinarism" to a natural system that would reflect genealogical relationships.

As we know from previous chapters, Miller was an admirer of Louis Agassiz, who in turn was a disciple of Georges Cuvier, founder of an influential school of comparative anatomy at the Paris Natural History Museum at the beginning of the nineteenth century. Indeed, Agassiz' "*Essay on Classification*" was characterized by one historian of science as a devote homage to the methods of classification expounded by Cuvier.[30] Although he did not live up to his own principles in practice[31], Cuvier always maintained – in theory at least – that animals had to be classified according to the degree of differentiation of organ systems judged by the investigator to be the most important ones for the survival of the organisms under investigation. Obviously, the brain was one of the most important organ systems for

---

[22]Panchen, 1992, ibid., p. 24; see also Mayr, E, 1982. The Growth of Biological Thought. The Belknap Press at Harvard University Pess, Cambridge, MA, p. 202; and n. 19.

[23]Ospovat, D. 1981. The Development of Darwin's Theory. Natural History, Natural Theology & Natural Selection (1838–1859). Cambridge University PPress, Cambridge, UK, pp. 101–113.

[24]Mayr, 1982, ibid., p. 846; Secord, 2000, ibid., p. 386.

[25]Secord, 2000, ibid. p. 430.

[26]Winsor, 1976, ibid., p. 141.

[27]Ospovat, 1981, ibid., p. 113.

[28]Panchen, 1992, ibid., p. 29.

[29]Winsor, 1976, ibid., p. 136.

[30]Lurie, E. 1960. Louis Agassiz: A Life in Science. The University of Chicago Press, Chicago, p. 205.

[31]Daudin, H., 1926, Cuvier et Lamarck. Les Classes Zoologiques et l'Idée de Serie Animale (1790–1830). Librairie Félix Alcan, Paris.

the classification of vertebrate animals, as was readily agreed upon by Chambers, Miller, and other authors. But this assessment of importance is made by the investigator prior to his investigation of natural diversity, and in the case of Chambers or Miller, it reflected deeply rooted preconceptions, such as their belief in progress. Miller did not care that an artificial system, one built on artificial kinds, might emerge from his reasoning, since the most important goal to him was that the system would reflect back on the First Cause. It is not the historical naturalness, but the logic of a classification that would reflect back on the rationality of the Creator. And since the power of rational thinking was thought of as the most important aspect that distinguishes humans from the animal kingdom, it seemed "natural" to choose the nervous system as the most important organ system used in the classification of animals. The problem with this approach is that different authors may deem different organ systems as the most important ones on which to build classifications, and classifications built on different organ systems may differ from one another. Red blood puts annelid worms closest to vertebrates according to Chambers; the nervous system puts squids closest to vertebrates. Which classification is the correct one, the natural one? Well, if classifications are allowed to be abstract constructs, then that question has no real bite, because there is no answer to it. In 1840, the highly influential British scientist and philosopher William Whewell outlined the problem in a nutshell: "The Maxim by which all Systems professing to be natural must be tested is this: – that the *arrangement obtained from one set of characters coincides with the arrangement obtained from another set*."[32] This is Whewell's appeal to the consilience of evidence, which we touched upon in the previous chapter already. It resonated a century later in the writings of another eminent philosopher of science, Carl Hempel: a "natural" classification is distinguished from an "artificial" one by the fact that "those characteristics of the elements which serve as criteria of membership in a given class are associated, universally or with a high probability, with more or less extensive clusters of other characteristics."[33] Replace the term "class" in this quotation from Hempel with the term "causally integrated system," and you obtain a roadmap for modern systematics.

6.5 William Whewell's "Consilience of the Evidence"

William Whewell opposed the theory of evolution, writing first against Chambers, and later opposing Darwin. But the method of scientific inference he outlined served Darwin[34] and ensuing generations of evolutionary biology very well indeed.

---

[32]Cited from Ruse, M. 1988. Philosophy of Biology Today. State University of New York Press, Albany, p. 54.

[33]Hempel, C.G. 1965. Aspects of Scientific Explanation, and Other Essays in the Philosophy of Science. Free Press, New York, p. 146.

[34]Stamos, D.N., 2007. Darwin and the Nature of Species. SUNY Press, Albany, NY, p. 96.

wolves and other dogs, let alone in pythons. But then, the tiger is not only a large cat but also a carnivore – carnivore now no longer understood as an ecological kind ("carnivores") but as a genealogical kind ("*Carnivora*"), a kind of mammals that is marked out by the relation of common ancestry, and for that reason includes tigers and wolves but excludes eagles and pythons. Thus, the tiger shares properties not only with other feline carnivores but also with all carnivorous mammals, characters that are absent in cows, horses, and antelopes, as also in non-mammalian carnivores. Then the tiger is also a mammal: it shares characters with other mammals – fur, single lower jaw bone, three ear ossicles, mammary glands in the females – that are absent in all animals that are not mammals. The tiger is an amniote, tetrapod, vertebrate, and so on. The individual tiger is thus a token of a natural kind, exemplified by the species *Panthera tigris*, but that natural kind is embedded in a whole hierarchy of natural kinds: Felidae, Carnivora, Mammalia, Amniota, Tetrapoda, Vertebrata, and this whole hierarchy of natural kinds is marked out by a series of characters from different organ systems at every one of its levels. So what can be said about the genealogical natural kind "tiger" is much richer than what can be said about the ecological natural kind "carnivores," and the reason is the "coming together" of a multitude of characters that mark out an entire hierarchy of natural kinds. Natural kinds exemplified by tigers or tetrapods are not as sharply demarcated as are the natural kinds exemplified by water or gold. Almost all water samples are composed of H_2O molecules, all gold nuggets are composed of atoms with the atomic number 79, yet there are tetrapods that lack four limbs in two pairs, such as snakes. This is where the counterfactual force of biological generalizations about genealogical kinds comes in degrees, because these kinds are not marked out by universal essential properties, but by some genealogically conditioned family resemblance. If I say: "If you were to come across a tetrapod tomorrow, it would have four legs," I would most likely be right. But in the unlikely event that the first tetrapod you stumbled across tomorrow is a snake, I would have been wrong. The tetrapod you stumbled across has lost its limbs as a consequence of evolutionary adaptation to life in a richly structured, complex habitat.[37]

If a tiger loses a leg in an accident, or lacks a leg as a consequence of a birth defect, it does not cease to be a tiger for that reason. Biological kinds such as "tiger" or "tetrapods" share a certain family resemblance, but one that nevertheless is causally grounded, namely in their common evolutionary origin. The common evolutionary origin is one of the properties marking out genealogical kinds.[38] An albino tiger born of earthly tiger parents still perfectly represents its kind. A tiger specially created on Mars that shares with earthling tigers all of their descriptive

[37]Chodrov, R.E., and C.R. Taylor. 1973. Energetic cost of limbless locomotion in snakes. Federation of American Societies for Experimental Biology, Proceedings, 32: 422. But see Walton, M., B.C. Jayne, and A.F. Bennett. 1990. The energetic cost of limbless locomotion. Science, 249: 524–527.
[38]Keller, R.A., Boyd, R.N., Wheeler, Q.D., 2003. The illogical basis of phylogenetic nomenclature. Botanical Reviews, 69: 93–110

properties, still is no example of the species *Panthera tigris*, because it does not share the evolutionary origin of that species.[39] The Martian tiger's properties are not rooted in the same causal process as are those of the earthling tigers. It is the coming together of a multitude of causally efficacious properties marking out a whole hierarchy of natural kinds, the occasional conflicting character distribution notwithstanding, which points to the naturalness of a classification. It is the many characters that snakes share with other reptiles, in particular with other scaly reptiles (squamates), that indicate that it is not the case that "snakes have no limbs," and for that reason should be classified with eels. What is instead the case is that eels have no limbs, because a multitude of characters from a multitude of organ systems indicate that eels are related to fishes that have no limbs and lungs, but fins and gills. In contrast, if the hierarchy of characters that marks out a hierarchy of natural kinds puts snakes with other scaly reptiles such as lizards, snakes do not have "no limbs," but instead they have "modified limbs," namely "lost limbs."[40]

This conclusion, based merely on the coming together of diverse characters from diverse organ systems, may still look suspicious to some skeptics. But one can go beyond Whewell's principle of consilience, which in systematics translates into character congruence. One can look into the causal grounding of those characters. The dissection of a python, or other basal snakes, will reveal hind limb rudiments. So there are intermediates, not snakes with four limbs in two pairs, yet also not snakes with no limbs, but snakes with hind limb rudiments. Useless for locomotion, these rudimentary hind limbs play an important role in the coordination of the mating behavior of boas and pythons, as the male uses them to trigger appropriate behavioral responses on the part of the female. If the mating was successful, the female python will deposit eggs, which the investigators can open to retrieve embryos at different stages of development. They will find that at an early stage of its development, the python embryo shows limb buds that announce the development of hind limbs that look closely similar to the limb buds of other tretrapods, such as a chicken, for example. As development continues, however, important cellular modifications occur in the limb buds of pythons, triggered by gene expressions that explain why the adult python has rudimentary hind limbs only.[41] The causal grounding of the character "loss of limbs," one of those marking out snakes as a natural kind, is then complete. This is the coming together of evidence in support of a natural classification, which really is a hierarchically structured system of natural kinds.

In his famous "*Essay on Classification*" from 1857, Louis Agassiz had used the fact of multiple limb reduction in scaly reptiles in an argument against theories of

---

[39]Hull, D.L. 1989. The Metaphysics of Evolution. State University of New York Press, Albany, NY.

[40]Platnick, N.I. 1979: Philosophy and the transformation of cladistics. Systematic Zoology, 28: 537–546.

[41]Cohn M. J., and C. Tickle. 1999. Developmental basis of limblessness and axial patterning in snakes. Nature, 399: 474–479.

descent with modification. He noticed in particular that one lizard family, the skinks, provided intermediate examples for all imaginable stages of limb reduction, going from four, to three, to two, to one toe, and on to the complete loss of external limbs. To him, this arrangement seemed to reveal a thoughtful concern for plenitude in nature, yet noting a mismatch or incongruence between the arrangement of species based on degree of limb reduction on the one hand, and their geographical distribution on the other, Agassiz concluded that "there is no connection between the combinations of their structural characters and their homes."[42] Lack of connection meant lack of causal connection, which in turn reveals freedom from laws of nature. For Agassiz, limb reduction in scaly reptiles merely "completed the scheme of nature" by rounding out the constructional type of "scaly reptile." Although (paradoxically) discoverable in nature, the "type" for Agassiz was the same abstract concept as was the class of cartilaginous fishes for Miller. For Agassiz, limb-reduced lizards did not point to an evolutionary bridge between lizards with well-developed limbs and snakes. Instead, limb-reduced lizards only rounded out the full "idea" of a "scaly reptile," thus revealing the unity of type as part of the plan of Creation. Lizards and snakes may seem to be quite different to the untrained eye, yet to the trained naturalist they both exemplify the same natural kind (i.e., they are both scaly reptiles or squamates), as is revealed by limb-reduced transitional forms.[43] One well-known lizard genus with limbs reduced to a functionless stage was named *Ophisaurus* by François Marie Daudin, in 1803.[44] His motivation for coining this name was to designate, by this name, a lizard that would almost perfectly bridge the gap between its tetrapod relatives and limbless snakes: half snake already (*ophi* from *ophis*, Greek, for snake), half still a lizard (*-saurus*). For Agassiz, the many lizard species showing graded degrees of limb reduction provided no indication that snakes might have originated from lizard-like ancestors who in the transition lost their limbs; the limb-reduced species for him merely filled in the gap between lizards and snakes, thus revealing continuity in the work of the Creator and the consequent plenitude in nature. Yet, contrasting such artificial with natural classifications, Darwin concluded: "Two hypotheses: fresh creations is mere assumption, it explains nothing further; points gained if any facts are connected."[45] Connecting the facts weaves together different lines of evidence; it delivers Whewell's consilience of the evidence. Noting the presence of rudimentary hind limbs in some basal snakes, Darwin was satisfied that "on my view of descent with

---

[42]Winsor, M.P. 2000. Agassiz's notions of a museum. The vision and the myth, pp. 249–271. In: Ghiselin, M.T., and A.E. Leviton (eds.), Cultures and Institutions of Natural History. Memoirs of the California Academy of Sciences no. 25. California Academy of Sciences, San Francisco, reference is to p. 263.

[43]Conrad, J. 2008. Phylogeny and systematics of Squamata (Reptilia) based on morphology. Bulletin of the American Museum of Natural History, 310: 1–182.

[44]Daudin, F.M. 1803. Histoire Naturelle, Générale et Particulière, des Reptiles. vol. 7, p. 346. F. Dufart, Paris

[45]DeBeer, 1960, ibid., part I, p. 53.

modification, the origin of rudimentary organs is simple."[46] Similarly, the example of anatomical correspondences such as intermediate degrees of limb rudimentation across scaly reptiles linking lizards with snakes within a group marked out by a distinct family resemblance, seems to well support Darwin's conclusion: "On my theory, unity of type is explained by unity of descent."[47]

---

[46]Darwin, Ch. 1859. On the Origin of Species. Charles Murray, London, pp. 450, 454.
[47]Darwin, 1859, p. 206.

Chapter 7
Respectable Science: What Is It?

Hugh Miller thought that his call for design, purpose, and goal-directedness in nature could be grounded in empirical science. He chastised Chambers for having drawn speculative conclusions that far transcended the empirical, i.e., observational basis. Chambers, in contrast, found regularity in nature expressed in the three-fold parallelism of the Great Chain of Being, embryonic development, and the Fossil Record, which he explained with his Law of Development. Darwin recognized the incompleteness of, and the consequent weakness of Chambers' system, and set out to 'connect the facts', to weave together all possible lines of evidence in support of his Law of Natural Selection. The astronomer Herschel belittled Darwin's theory as the "law of the higgledy-piggledy". Today, we hear calls for 'Creation Science', and 'Intelligent Design' is propagated as scientific, not by evolutionary biologists, but by scientists nevertheless. So what is science and what isn't?

A definition of science has proven largely elusive. To understand what science is, one must observe how its practitioners do science. This is how practicing scientists become the fruit-flies of philosophers of science. Scientists enter into causal relations with the physical world in order to better understand it. Philosophers of science observe scientists and write books about how scientists practice their profession. The four most popular philosophers of science of the 20th Century – popular both with scientists and with the broader public – created a rather pessimistic view of science. For Karl Popper, there is no way to know a scientific theory – which he defined as a universal law of nature – to be true. We can only know it to be false, if it fails a crucial test. For Thomas Kuhn, scientists themselves create large parts of the world they are investigating through their scientific theories. As theories change, the relevant parts of the world change with them. Knowledge of the world thus remains always relative to the way scientific theories describe the world. Imre Lakatos sought a synthesis of the views of Popper and Kuhn. For him, science cycles through progressive and degenerating research programs. Paul Feyerabend took the Kuhnian relativism to its logical conclusion, reducing science to its social and political dimensions in an ever-changing historical context. Whether 'Creation Science' exists or not is no longer a matter of a proper definition of science, but a question of whether its proponents can organize themselves in a socially and politically effective way.

Such pessimistic views of science are rooted in the application of certain concepts from the philosophy of language to the interpretation of the history of science, mostly physics. But these concepts have proven inadequate, or incomplete, if used in an attempt to understand the language of science and its historical change. New approaches have been developed that emphasize not the merely descriptive function of scientific theories, but more importantly their proper grounding in causal relations. Successful scientific theories are theories which reliably, and hence predictably, link causes to their effect(s). This allows for a far more optimistic view of scientific progress, but bans the notions of design, purpose,

O. Rieppel, *Evolutionary Theory and the Creation Controversy*,
DOI 10.1007/978-3-642-14896-5_7, © Springer-Verlag Berlin Heidelberg 2011

and goal-directedness from natural science. For those concepts to survive in biology it
would be necessary to show how they can be grounded in the natural causal propensities
and dispositions of naturally occurring entities such as organisms, or species.

7.1   The Definition of Science

For Chambers, science was all about natural laws. But it was also about integration, grand visions, bold theories, and simplicity. What he proposed was a cosmology of a developing universe, which explained change in the inanimate animate departments of the earth, on the basis of only two universal laws of nature: the Law of Gravity and the Law of Development, the two running parallel relative to one another. A unified vision of the universe – two laws and two only to which all phenomena of the physical and of the living world could ultimately be reduced.

Miller would not take issue with the quest for natural laws as the primary goal of science. But he did take issue with what he considered to be unfounded sweeping generalizations presented by the author of the *"Vestiges."* According to him, scientific generalizations are not a matter of fertile imagination, but of unbiased observation. To him, cosmology, geology, anatomy, and paleontology were all separate departments of science, each tied to different methods, different equipment, and different questions to ask and answer. Scientists were experts in their respective fields, who respected the limits of the scope of generalizations that can be made on the basis of all available observed facts. But he, furthermore, found natural science as a whole to be constrained by certain limitations. The Law of Gravity could explain the formation of sedimentary rocks, the formation of sedimentary rocks could explain the succession of fossils, but the succession of fossils could not explain the origin of ethical and moral norms of human social behavior.

According to Darwin, "The grand question which every naturalist ought to have before him when dissecting a whale, or classifying a mite, a grampus or an insect is What are the Laws of Life?"[1] As we have seen, the Laws of Life understood in the way Darwin explained them to us are different from the Laws of Physics. But still, and more generally, to discover regularities in nature, to understand them as the expression of lawfulness in nature, to formulate such natural laws preferably in the rigorous formal language of mathematics, and to test these laws were and still are commonly understood to be the business of science. But there were, and still are, different accounts of what science is, or should be. Chambers sought bold generalizations; Miller wanted a science built on pure, unbiased, theory-free (i.e., "raw") observation. Only such science would truly reflect the causal structure of the world, and provide objective knowledge. But since Miller's time, there have been philosophers of science who, on very strong grounds, have argued that theory-free

---

[1]DeBeer, G. 1960. Darwin's notebooks on transmutation of species, part I. Bulletin of the British Museum (Natural History), Historical, 2:2: 69.

observation is impossible. One of the first to do so was, indeed, Sir Karl Popper. So what is science, what is scientific, and what is not? Can a science of biology incorporate design and purpose, and if not, why not? Can a theory of species transformation be governed by universal laws of nature, and if not, why not?

On November 9, 2005, the American Association for the Advancement of Science (AAAS) released a statement from its CEO, Alan I. Leshner that read: "By definition, scientific explanations are limited to rigorous, testable explanations of the natural world and cannot go beyond."[2] This is a common-sense view of science that goes a long way, one that would be acceptable to the great majority, if not all scientists. It is also a statement that reflects the practice of science: to seek the best explanation of the causal structure of the world, and to formulate these explanations in terms of rigorously testable theories. But looked at a bit more closely, this statement turns out to be surprisingly complex. First, it claims that science can be defined (i.e., a distinction can be drawn between what is proper science and what is not on the basis of a definition). Second, it invokes the issue of scientific explanations and their rigorous test: what are "rigorous scientific explanations," how can we test them, and how will we interpret the results of such tests? Is Chambers explanation of the order in nature as being the result of the action of the "Law of Development" a rigorous one, or one that can be tested rigorously? Can such a test of the "Law of Development," or of species transformation quite generally, be based on observation? And what does it mean to test a scientific theory if it should turn out that observation itself is never theory-free?

The demarcation of science from everything that is not science, usually called metaphysics, has vexed many minds, philosophical and scientific. Many solutions have been proposed, some of them will be discussed later. However, at a very basic level, some philosophers of science, as well as some scientists, have argued that science, as anything else, cannot really be *defined*. This is because definitions establish sameness of meaning, or synonymy, between words and nothing more. Definitions are not made true by observation; they are made true by the collective use of the language in which definitions are given. Definitions, consequently, say nothing about the world, they only fix the use of words. One cannot define the world, but one can define the meaning of words used to talk about the world. One such definition would be: "All bachelors are unmarried men." Through this definition the term "bachelor" becomes synonymous with the term "unmarried man." According to the definition, "bachelor" and "unmarried man" mean exactly the same thing – but that definition does not also tell us what a "man" is and what "unmarried" means. New definitions appear to be called for to explain the meanings of the words that were used in the first definition, such as "man" and "bachelor." But then, additional definitions will be required to define the words used in the definitions that give the meaning of the words "bachelor" and "man," and so on. Definitions will therefore ultimately always run out of words. Attempts to rigorously define terms, some philosophers argue or have argued, will ultimately lead to

---

[2]http://www.aaas.org/news/releases/2005/1109kansas.shtml.

the edge of language where no words are left, such that the meanings of words have to be established by ostensive gestures. Pointing at an acquaintance over there, one might say: "This is a bachelor!" – "What do you mean?" might likely be the interlocutor's reaction. Ostension, again, does not seem to be able to "step out of language."[3] The solution would seem to be not to define words, but to find out how words are used in a linguistic community, how English speakers use the terms "bachelor" and "unmarried man" relative to the world of time, space, and people. In parallel, one could argue that science is so multifaceted that it cannot be defined in any philosophically stringent sense of "definition," but that one must look at how science is practiced by the scientific community to understand what science is. We can define scientific theories as those theories that are rigorously testable, but that only generates the task of defining "rigorous," "theory," or "test." Alternatively, to learn how science works, one can observe what practicing scientists call a testable theory, and how they go about to test it.

7.2   Ludwig Wittgenstein and the "Verification Principle"

Miller not only believed in the possibility to define special branches of science, but furthermore appealed to theory-free observation, claiming that it forms the basis of all objective scientific knowledge. He claimed species transformation to be idle speculation because it cannot be directly observed in the Fossil Record. With his arguments, Miller looked back on the philosophy of British empiricists such as John Locke (1632–1704), David Hume (1711–1776), and John Stuart Mill (1806–1873), who founded the philosophical tradition called *Empiricism*. This tradition witnessed a major revival in the early twentieth century when it was picked up and further developed by Bertrand Russell and Ludwig Wittgenstein. In the 1920s, a group of philosophers, mathematicians, and scientists from the University of Vienna began to organize regular weekly meetings to discuss and further develop empiricist philosophy.[4] It was a rather exclusive and very active circle, including some of the most brilliant faculty members of the University of Vienna, who sought contact and exchange of ideas with fellow philosophers and scientists from all over Europe, Great Britain, and the United States. They invited colleagues to their Thursday evening meetings, participated in the organization of international symposia, and founded a society, called the "*Verein Ernst Mach*," to gain public influence in adult education and Austrian politics. Members of the Vienna Circle were among the first to seek a special philosophy of science, a philosophy that would explain the logic of scientific discovery. The focus of their interest was the

---

[3]Wittgenstein, cited in Oberdan, Th. 1993. Protocols, Truth, and Convention. Rodopi, Amsterdam, p. 106.

[4]Stadler, F., 1997. Studien zum Wiener Kreis. Ursprung, Entwicklung und Wirkung des logischen Empirismus im Kontext. Suhrkamp, Frankfurt a.M. See also Janik, A., and S. Toulmin. 1973. Wittgenstein's Vienna. Simon & Schuster, New York.

logic of the language of science, the logic of scientific theories, and how these relate to observation reports, indeed, the logic of verification and falsification of scientific theories. But the first task at hand surely was to identify what kind of a statement would qualify as a genuine scientific theory, as opposed to statements that had to be rejected as nonscientific (i.e., metaphysical). The first task identified by the members of the Vienna Circle therefore was the demarcation of science from metaphysics.

Ludwig Wittgenstein[5] is claimed by many to have been the most important philosopher of the twentieth century. He was the eighth child born into a prominent and wealthy family in Vienna on April 26, 1889. He grew up in a posh environment that fostered intellectual achievement and exposed him to high culture. During his early academic studies at the high school and at university level, Wittgenstein was captured by physics and engineering. His interests for the foundations of mathematics eventually led him to the study of philosophy. Following the advice from Gottlob Frege, who taught logic, philosophy of language, and philosophy of mathematics at the University of Jena in Germany, Wittgenstein enrolled as a student of Bertrand Russell at Trinity College of Cambridge University, England. At the outbreak of World War I, Wittgenstein returned to his native country and volunteered to serve in the Austro-Hungarian army. During the war, and the time he spent as a POW in Italy, Wittgenstein wrote the only book-length manuscript that he ever published during his life, the famous "*Tractatus Logico-Philosophicus.*" The book is a difficult read, and Wittgenstein found it equally difficult to find a publisher, but once it appeared, in 1921 in German, and in 1923 in an English translation prefaced by Bertrand Russell, it was to revolutionize philosophy. Wittgenstein himself thought that his book had solved all the fundamental problems of philosophy worth considering. After the war, from which he emerged as a decorated hero, Wittgenstein returned to Austria to pursue several nonphilosophical interests and vocations in and around Vienna. It was during that time that Wittgenstein came into contact with members of the Vienna Circle, who in their meetings debated the *Tractatus*, tediously studying the book line by line. In 1929, Wittgenstein eventually returned to Cambridge University, where he became a professor known for his idiosyncrasies, his hot temper, and his biting intellect.[6] Publishing little, he nevertheless exerted an enormous influence in philosophy until his death on April 29, 1951.

Claiming influence from the younger Wittgenstein, the members of the Vienna Circle thought they had found an elegant and easy way to demarcate science from metaphysics, namely through the "Verification Principle." And that principle states that only those statements that can potentially be verified should count as scientific; all others should be discounted as metaphysical or indeed as nonsense. That was an unnecessarily harsh position to take *vis-à-vis* metaphysics, one that also faltered over time. What the verification principle implied was that for anyone to make a

[5]Monk, R. 1991. Ludwig Wittgenstein: The Duty of Genius. Penguin, London.
[6]Eidinow, J., and D. Edmonds. 2005. Wittgenstein's Poker. Faber and Faber, London.

meaningful statement, the speaker must be able to specify under which conditions that statement could be shown to be either true or false. If the speaker could not specify such conditions, she would not know the meaning of her statement, and hence talk nonsense. The possibility of verification (i.e., possibility to specify the conditions under which a statement could be found to be either true or false) was what in the eyes of those philosophers gives a statement its meaning. Here is Alfred J. Ayer's empiricist battle cry: "We say that a statement is factually significant to any given person, if, and only if... he knows what observations would lead him, under certain conditions, to accept the proposition as being true, or reject it as being false."[7] In more general terms, the verification principle admits only testable statements as scientific, that is, statements for which conditions could be sketched that would show them to be true or false.[8] This does not mean that a statement (i.e., a theory) *must* actually be shown to be true or false. It only means that conditions have to be specified under which a theory *could* be shown to be true or false. Quite generally, then, scientific statements are characterized by their testability, where a test is supposed to show a scientific statement to be right, or probably right, or wrong, or probably wrong relative to the observable world of experience.

A "statement" in the jargon of philosophy is not just any kind of sentence; it is a sentence that makes a proposition about the world (i.e., one that proposes something to be the case in the world of experience). "This apple is red," or "all ravens are black," are statements that propose something to be the case in nature. Laws of logic are not true or false because of the way the world is, whereas statements about the world of space, time, and matter are true or false because of the way the world is. For example, one of the very basic laws of logic we already encountered, the law of noncontradiction, specifies that a proposition "P" and its negation "not-P" cannot both be true ([P and not-P] is false). Of two contradictory propositions, P and not-P, only one can possibly be true, the other must be false. This law holds no matter what the propositions say about the world. We do not need to look at an apple, and we, in fact, do not even need to know what an apple is in order to know that the following statement can under no circumstances be a true statement: "this apple is all-over red and this very same apple is not all-over read." We know a priori, on the basis of pure reflection (i.e., as a matter of pure logic without any backing of observation) that any statement of the form "P and not-P" is false. Propositions about the world, in contrast, are true or false depending on what condition the world is in. If, when pointing to an apple, somebody states "this apple is red," the statement will be true only if it is confirmed by observation; if the observation results in a report of a green apple, the statement will be false. "All ravens are black" will be false if a white raven can be pointed to or reliably reported on. Importantly, therefore, we cannot know the statement that "all ravens are black" to be true or false a priori, that is, on the basis of pure reflection. This would be possible only if we *defined* ravens as

---

[7]Ayer, A.J. 1952 [1946]. Language, Truth & Logic. Dover, New York, p. 35.
[8]Godfrey-Smith, P. 2003. Theory and Reality. An Introduction to the Philosophy of Science. The University of Chicago Press, Chicago.

being black – but that would mean to define the word "raven," rather than confirm or disconfirm the empirical fact that "all ravens are black." Whether "all ravens are black" is true or false, we can only know a posteriori (i.e., after we investigated the world) looking for a possible white raven.

So, for logical empiricists, to give the meaning of a statement that says something about the world – as empirical scientific statements would have to – means to specify the conditions under which the statement could be shown to be either true or false; it means to specify the conditions under which a statement would be testable.[9] Statements for which such conditions could not be specified were rejected as metaphysical by the logical empiricists. That does not mean that anybody would have to abandon beliefs such as, for example, the belief in a supranatural agent residing beyond space–time and directing the natural course of events toward a goal and purpose. It only means that for as long as we cannot specify the conditions under which such propositions about supra-natural agents could be shown to be true, or false, such propositions cannot count as scientific. The same is true for the concept of perfect adaptation. It is surely possible to stipulate that the eye, or the bacterial flagellum, is such a complex structure that it could not have evolved through variation and natural selection, but must have been intelligently designed. Again, unless conditions can be specified under which such a statement of "intelligent design" can at least potentially been shown to be true or false, such a statement cannot be claimed to be scientific.

7.3   The Illusion of "Theory-Free Observation"

All that may sound very acceptable in theory, even in scientific practice, but it caught the logical empiricists in a bad corner. Logic, as also mathematics, has a certain beauty and charm to it: given the right axioms, we know that "$2 + 2 = 4$" is true, and that "$13 - 6 = 3$" is false. There is not much to argue here (although at least one philosopher voiced strong doubts about such apparently easy truths and falsehoods, again appealing to Wittgenstein as his patron saint). Not so with propositions whose truth or falsehood depends on the way the world is; for to specify the conditions under which a scientific statement could be shown to be true or false requires the possibility to relate that statement to observation reports, and in contrast to mathematical theorems, observation reports are less easily agreed upon. It is well known how difficult it is to get accurate eye-witnesses' reports on crime scenes or traffic accidents. Here is where Miller's empiricist call for theory-free, untainted, "raw" observation becomes crucial, and controversial. It was Moritz Schlick, Professor for the History and Theory of Inductive Sciences at the University of Vienna, who initiated the Thursday evening gatherings that resulted in the formation of the Vienna Circle. At these meetings, Schlick continued to defend the

---

[9]Godfrey-Smith, 2003, ibid., p. 27.

doctrine of theory-free observation even after some of the attendees had realized how difficult, indeed impossible, the defense of that doctrine is. Schlick famously exclaimed: "I have been accused of maintaining that statements can be compared with facts. I plead guilty. . . I have often compared propositions to facts. . . I found, for instance, in my Baedeker the statement: 'This cathedral has two spires', I was able to compare it with 'reality' by looking at the cathedral, and this comparison convinced me that Baedeker's assertion was true."[10] The issue of what was meant by "facts" aside, the question arises whether Schlick's observation of a cathedral with two spires could, indeed, be theory-free. Some members of the Vienna Circle had come to realize that this was not, in fact, possible. Things are not as easy as Schlick tried to portray them.

It was an outsider who had drawn their attention to this problem. Not on the faculty of the University of Vienna but a school-teacher instead at the outset of his career, Karl R. Popper nevertheless had high hopes to be invited to become a member in Schlick's Circle, and while he maintained close contact with some of its participants, he was never admitted as a regular member to the "core" of the Circle. Popper remained a member in what has been called "the periphery" of the Circle[11], a fate he shared with Wittgenstein, who later became his bête noire in philosophical debate.[12] However, Schlick recognized the strength of Popper's philosophy and arranged for the publication of his manuscript on the "*Logic of Scientific Discovery*" (1935) in a book series edited by the Vienna Circle. It was this book that prepared the ground for Popper's ascent to superstar status amongst philosophers of science through a long career, which he ended as Sir Karl at the London School of Economics.

Popper shared the logical empiricist concern for the demarcation of science from metaphysics, but he was far less enthusiastic about the verification principle. His skepticism was motivated, in part, by his insight that theory-free observation is impossible. To observe the two spires of a cathedral might seem as straightforward as to observe the hands move on the clock mounted on the bell-tower, or as taking a reading of an instrument in a physicist's laboratory. But Popper argued that each time when we take a reading from an instrument we "rely on the hypotheses of geometrical optics, on the theory of solid bodies, on the correctness of Euclidean Geometry in small space, on the hypothesis of the existence of things, and innumerable other hypotheses."[13] Several years later, Norwood Hanson[14] provided another famous example for the theory-ladenness of all observation: imagine a doctor seeing a patient at his desk. So as not to forget to order a replacement, the

---

[10]Cited from Oberdan, T. 1993. Protocols, Truth, and Convention. Rodopi, Amsterdam, p. 61.

[11]Stadler, F. 2007. The Vienna Circle. Context, Profile, and Development. In: Richardson, A., and T. Uebel (Eds.), The Cambridge Companion to Logical Empiricism. Cambridge University Press, Cambridge, UK, pp. 13–40.

[12]Eidinow and Edmonds, 2005, ibid.

[13]Popper, K.R. 1979. Die beiden Grundprobleme der Erkenntnistheorie. J.C.B. Mohr (Paul Siebeck), Tübingen, p. 391.

[14]Hanson, M.R. 1958. Patterns of Discovery. Cambridge University Press, Cambridge, UK.

doctor had earlier in the day put an expired X-ray tube on his desk. It could seemingly be argued that the doctor and the patient *see* the same object on the doctor's desk. However, the patient who is blissfully ignorant of X-ray technology does not *see* the expired X-ray tube *as* the same thing as the doctor does, who is worried about the cost of its replacement. Indeed, the patient might look at this strange devise wondering what, in fact, it is? It is the context from within which we are looking at something that determines not what we see, but what we recognize when looking at something. We can look and yet see nothing. We can see something and then see the same thing *as* something else; it happens when the doctor recognizes the puzzled expression in his patient's face and then explains to his/her what he/she is looking at.

Commenting on examples such as these, the eminent Harvard philosopher, Willard van Orman Quine, once a visiting attendee at the Vienna Circle sessions, concluded: "The notion of observation as the impartial and objective source of evidence for science is bankrupt."[15] But if that is so, if there cannot be any theory-free observation statements, then an observational report can never be fully verified. Yet if an observation statement cannot be known to be true (in the strong, indeed ultimately timeless sense of the concept of "truth"), it also cannot conclusively verify or falsify a scientific theory. Following Popper's insight, the verification principle had to be dropped, or at least revised. At least some members of the Vienna Circle, most notably those around the leading philosophers Otto Neurath and Rudolf Carnap, learnt that lesson fast, and turned to the development of ways to judge scientific theories in terms of degrees of confirmation or degrees of disconfirmation (i.e., on probabilistic grounds). On that account, a scientific theory is not either true or false, and nothing in between. A theory is more or less probable, depending on the degree to which is it supported by the available evidence. In contrast, Popper took a radical skeptical turn.

7.4 Karl Popper's Demarcation of Science from Metaphysics

Karl Popper is probably the one philosopher of science of the twentieth century who found the broadest reception and acclaim not only throughout society, but, in particular, also throughout the scientific community. In spite of his utterly skeptical outlook, Popper's views on what science is and how it works deeply influenced the public perception of science. Some renowned scientists, such as the biologist Sir Peter Medawar, declared Popper "the greatest philosopher of science ever."[16] Popper, indeed, had much more success with scientists and the broader public, than with his fellow philosophers of science. One of the reasons may be Popper's

---

[15]Quine, W.V.O. 2000. Epistemology naturalized, pp. 292–300. In: Sosa, E., and J. Kim (Eds.), Epistemology, An Anthology. Blackwell, Malden, MA, (reference is to p. 299).

[16]Magee, B. 1973. Popper. Fontana, New York, p. 9.

accessible prose: Popper referred to himself as a "common sense philosopher" and argued that "common sense" should form the starting point for philosophy.[17] A weak reading of Popper engenders among scientists a critical attitude toward their own pet theories: in that sense, his philosophy encourages a culture of critical discussion, to seek flaws and mistakes in one's own scientific work rather than ultimate confirmation.[18] Things get more serious, however, when we "look upon the critical attitude as characteristic of the rational attitude," where rationality is the capacity of logical thinking, and where Popper recognized "deductive logic as the organon of criticism."[19] For here we cross the bridge from "common sense" philosophy to a full-blown theory of knowledge. But philosophy of science was not the only branch of philosophy to which Popper contributed. Although funding agencies for science still require grant proposals to sport the Popperian jargon of testability and falsifiability, fellow philosophers tend to hold him in higher regard for his social and political writings.[20] Why should that be? The reason is that following Popper's position on science all the way home, it turns out in the end to be untenable. Popper wrote on the *"Logic of Scientific Discovery,"* but when asked what he would consider a solid piece of scientific knowledge to be, his answer was that such knowledge cannot exist. How come? What about the calculations that landed people on the moon: were they pure lucky guesswork, or were they based on solid scientific knowledge? There is an easy and a more complicated way to answer these questions in Popperian terms.

The easy way goes somewhat like this: call a "scientific statement" a theory. The theory is scientific if it can be tested, and – according to Popper – at least potentially falsified. Falsification occurs when an observation statement contradicts the theory. Such contradiction ($[P$ and not-$P]$ is false) occurs in logical space, and not in the world. The contents of the world, water and gold, tigers and elm trees, cannot contradict each other. Only our statements about the world can be contradictory. So what is required for a theory to be falsified is that one statement about the world (i.e., the theory) is contradicted by another statement about the world (i.e., an observation statement). A theory is logically falsified if it clashes with an observation statement. But to render a theory not true, the contradictory observation statement itself must be true. But this, according to Popper, we will never know, because all observation is theory-laden. If all observation statements are theory-laden, then they can neither conclusively verify, nor conclusively falsify a theory in the logical sense. Popper's philosophy of science leaves the logic of scientific discovery stuck in the mud.

---

[17]Popper, K.R. 1973. Objective Knowledge. An Evolutionary Approach. Oxford University Press, Oxford.

[18]For an example see Patterson, C. 1982. Classes and cladists or individuals and evolution. Systematic Zoology, 31: 284–286.

[19]Popper, K.R. 1973, ibid., p. 31.

[20]Newton-Smith, W.H. 2005. Karl Popper (1902–1994), pp. 110–116. In: Martinich, A.P., and D. Sosa (Eds.), A Companion to Analytic Philosophy. Blackwell, London.

A more complicated, but also much more sophisticated, approach to Popper's philosophy runs along the following lines. Popper was an ambitious man, so he wanted to grasp the nature of science to its full extent. He would not bother about minor scientific achievements. In analyzing the logic of scientific discovery, he set the bar of scientific knowledge as high as possible. The scientists he chose to look at were Newton and Einstein, who had captured the workings of the universe in the form of universal laws of nature. Universal laws were what Chambers and Miller argued about, and universal laws were what Popper wrote about when he developed his falsificationist philosophy of science.

As already mentioned, universal laws are true, or false, everywhere, all the time, throughout all possible worlds, given the relevant conditions for them to obtain at all. Universal laws impart necessity on the natural course of events. If a universal law is true, the world could not be any other way than the way it is described and explained by that law. Popper called "universal laws" scientific theories, or rather the other way around: he defined scientific theories as universal laws, or universal statements, which says the same.[21] He further asked the question: is it ever possible to know whether such a theory is true? His answer was negative. The reason is that because such universal theories hold in all possible worlds at all times, it would require an infinite number of its positive instantiations (i.e., an infinite number of tests with positive outcome), for us to know that it is true. But an *infinite* number of tests, positive or negative, cannot ever be achieved – not just in practice, but also in principle. So we will never know whether such a theory is true. This is step one into Popper's skepticism.

Step two is the following: although it can never be known whether a scientific theory (defined as a universal statement) is true, it can be known that it is false. How so? Well, Popper was an admirer of Darwin and his theory of evolution. According to Darwin's theory of evolution, and in its simplest terms, organisms vary, and selection chooses from those variants those that are fitter than others. Popper thought that scientists obtain knowledge in a similar way, through variation and critical selection of theories. Since we cannot ever know any theory to be true, it does not really matter how we obtain a theory in the first place. It might be a hunch, some intuition, guesswork, good luck, or whatever: just try a theory by testing it! That is, the scientist issues a universal statement and then proceeds to test it, not in an attempt to prove it right, but rather in an attempt to prove it wrong. Popper of course had an agenda when he defined scientific theories as universal statements. The reason is not only high ambitions for science, but also the logical properties of universal statements. As already pointed out, the issue at stake is that universal statements allow the deduction of forbidding singular statements. A by now well-worn prototype of a scientific theory, understood as a universal statement, is "All ravens are black." An example for a deductively related forbidding singular statement is "There is no white raven here now." We can now build a logical, (i.e.) a deductive

[21]Stamos, D.N. 2007. Popper, laws, and the exclusion of biology from genuine science. Acta Biotheoretica, 55: 357–375.

bridge between the theory and the state of affairs it forbids. If it is true that "all ravens are black," then it must also be true that "all things that are not black are not ravens." But if it is true that "there is nothing that is not black and a raven," then it is also true that "no white thing is a raven." In that way, it is possible to deduce from the theory "all ravens are black" the forbidding statement "there is no white raven here now." It is at this point that the bird watchers can go out and explore the world. No matter how many black ravens they would have seen, they cannot show the theory to be true, because the number of black ravens that have been observed will always remain a finite number, which will never be sufficient to verify a universal statement. But if it so happens that an ornithologist is spotting a white bird, which he/she identifies as a raven, he/she can exclaim while pointing at it: "there is a white raven right here and now"! This observation statement logically contradicts the forbidding statement that was deductively derived from the theory, and hence it logically falsifies the theory that "all ravens are black." Popper maintained that scientists learn from their mistakes. The task of scientist is to propose theories, the bolder the better – and what bolder theories could there be than universal statements, true or false everywhere and at all times? Proposed theories are then to be tested as severely as possible, not with the aim to verify them, since this is impossible, but with the aim to falsify them. It is the falsification of theories that renders those obsolete. If falsification occurs, the scientist has to drop the old theory, and come up with a new one, which he/she then proceeds to test, and so on. For Popper, conjecture and refutation are the mechanisms that push science forward on its road to increasing success, and not positive knowledge about the world. Konrad Lorenz, the Nobel Prize winning student of animal behavior, characterized Popper's evolutionary approach to the growth of knowledge in an intuitively appealing way. If a simple creature such as a paramecium encounters an obstacle in its path, it first retreats, then moves in a randomly chosen different direction, thus showing that it has "learnt" something about its world. It may not "know" where to move next, but it "knows" it cannot move straight on.[22] Metaphorically speaking, the same holds for the scientist.

But not so fast! Was it not Popper's first claim to fame to have recognized that all observation reports are theory-laden? If observation reports are always theory-laden, they can never be known to be true. If, for that reason they cannot verify a theory, then, for that very same reason, they also cannot falsify a theory. Popper's reasoning seems caught in a corner. Look at the observation statement, "there is a white raven here now." It requires that the animal in question be white, not blite, nor of a shade of grey that could still pass as black. But worse, it requires that the animal in question has correctly been identified as a raven. But to identify any animal (or plant) as being of a certain species is very heavily theory-laden. First, it requires a theory of species; second, a theory of species identification; and third, in the case of the white raven, expert knowledge in ornithology. The theory that "All swans are white" was supposedly falsified when black swans were discovered in Australia, or

---

[22]Lorenz, K. 1973. Die Rückseite des Spiegels. Piper, Munich, p. 16.

so it seems. If fact, it was not, for the European white swan and the Australian black swan were found to belong to two different species.

Popper avoided this trap by drawing a distinction between the logical and the practical falsification of a theory. If the theory is "all swans are white," and the observation report is "there is a black swan here now," then the theory is *logically* falsified, and there is nothing that can be done about it. But this is a matter of the laws of logic, it is not a matter of the living world. The theory is falsified in logical space, but it may not be false in the living world, where black swans are found to live in Down Under – who would have guessed? So according to Popper, a theory can be technically, that is, logically, falsified, but it does not, at the same time, need to be empirically false (i.e., false relative to the experienced world). But let us go back to the theory that "all ravens are black," for in this case, no new species of white ravens has (yet) been described. Our theory therefore is "all ravens are black," yet a competent birdwatcher on the shores of Lake Geneva spots a bird of which he says: "there is a white raven here now." Is the theory falsified? Only if the bird pointed at is truly white and truly is a raven. But how could we know? According to Popper, all observation statements are theory laden, and theories can never be known to be true. So, for the empirical – not merely logical – falsification of the theory that "all ravens are black" to occur, it is required that the entire community of competent bird-watchers unanimously accepts the fact that there is, indeed, a white raven here now. That is, the community of all competent experts of the time has to *accept* an observation statement as true, or at least as provisionally true, for it to function as a falsifier, or at least as a provisional falsifier of a theory. However, this has neither anything to do with logic nor with objective knowledge. Instead, it has everything to do with scientific expertise that is claimed to justify scientific authority. It is the "accredited observers of the time," as members of the Vienna Circle put it, who decide which observations are true or false, not nature.

Popper compared the scientific experts of the time, who accept or reject observation statements, to "something like a scientific jury."[23] Emphasizing the importance of the verdict of the jury, Popper elucidated the term "verdict" as derived from the Latin expression *vere dictum*, which means "truly spoken" – in answer to a fact. But given his emphasis on the theory-ladenness of all observation statements, it seems that no such truth can be had, for which reasons his critics argued that such ruling by a jury turns science into a matter of mob psychology.[24] Philosophical polemics to the side, what *do* we know at the end of scientific inquiry? "Precious little" is Popper's answer, one that is motivated by his radical skepticism. Calling Popper's science a matter of mob psychology is certainly going too far, but Popper did call science a "game." In his "*Logic of Scientific Discovery*," Popper wrote: "The game of science is, in principle, without end. He who decides one day that

---

[23]Popper K.R. 1974. Replies to my Critics, pp. 961–1197. In: Schilpp P.A. (ed.), The Philosophy of Karl Popper, vol. 2. Open Court, La Salle, IL. p. 1111.

[24]Newton-Smith, W.H. 1981 [1994]. The Rationality of Science. Routledge, London, p. 64.

scientific statements do not call for any further test, and that they can be regarded as finally verified, retires from the game."[25] He wrote that the bold structure of the theories of science rises above a swamp. Science is like a building erected on piles that are driven from above into the swamp, without ever reaching natural bedrock. Popper concluded his famous book with the observation that "Science is not a system of certain, or well-established statements... it can never claim to have attained truth, or even a substitute for it, such as probability."[26] As is documented by the many appendices which he later added to "The Logic of Scientific Discovery," developments in modern physics, especially the rise of quantum mechanics, forced Popper to deal with issues of probability theory more than he ever would have liked to.

7.5   Thomas S. Kuhn and the Social Nature of Science

Another philosopher who was accused of turning science into a game, or worse, a matter of mob psychology, was Thomas S. Kuhn.[27] After studying physics, Kuhn went on to teach history and philosophy at Harvard, at the University of California in Berkeley, at the Massachusetts Institute of Technology, and at Princeton University. While in Berkeley, he wrote his most famous book, "The Structure of Scientific Revolutions" (1962). It was an immense success, picked up not only by fellow philosophers, sociologists, and historians of science, but also by scientists and the broader public. His is likely the most successful book ever published in the philosophy of science, in terms of the number of copies sold, the number of languages into which it has been translated, and the number of times it was cited in secondary sources. A landmark event when it was first published, it became highly influential in the years thereafter, yet Kuhn spent much of his later life defending his views against what he believed to be erroneous interpretations of his book by commentators. However, there is some irony surrounding the publication of this book, which has been characterized as "something of a 'Trojan horse' situation."[28]

The Trojan horse is the ingenious device with which Odysseus along with Greek soldiers managed to enter the beleaguered city of Troj, all hiding inside the immense wooden construction. The episode marked the end of the Trojan War. In a symbolically similar way, it was the fathers of logical empiricism, former members of the Vienna Circle, who arranged for the publication of Kuhn's "Structure" in their book series on the "Unity of Science," which, once it was published, contributed in major ways to the demise of logical empiricism and the philosophy

---

[25]Popper, K.R. 1959. The Logic of Scientific Discovery. Hutchinson, London, p. 53.
[26]Popper, 1959, ibid., p. 278.
[27]Grandy, R. 2006. Thomas S. Kuhn (1922–1996), pp. 371–377. In: Martinich, A.P., and D. Sosa. A Companion to Analytic Philosophy. Blackwell, Malden, MA.
[28]Godfrey-Smith, 2003, ibid., p. 75.

of science that was associated with it. When Hitler came to power in Germany in 1933, and fascism started to raise its ugly head ever higher in Austrian universities, the members of the Vienna circle realized it was time to go. The same was true for the members of a similar philosophical circle that had formed in Berlin, and that had established close ties with the Vienna Circle throughout the 1920s. The leading members of the Berlin *Society for Empirical* or *Scientific Philosophy* shared a number of important interests with the members of the Vienna Circle, one of which was the unification of science. It is important to realize that these empiricist philosophers saw science embedded in the social and political context of its time. Hence the interest of the Vienna Circle's public arm, the *Ernst Mach Verein*, in adult education and politics. Science was to provide important guidance in the making of policies governing the cultural, social, and political domains of society. To do so most efficiently, it would benefit science if it could be presented to the broader public as a unified endeavor. At the same time, science would strengthen its own philosophical foundations as a consequence of such unification. It was the sociologist Otto Neurath, a member of the Vienna Circle, who founded the *Unity of Science Movement* to carry these goals forward, which ultimately resulted in an empiricist attitude toward all sciences, and the search for a single comprehensive scientific language that is firmly rooted in formal logic. Scientific theories of all kinds would have to be logically structured systems, which would ultimately have to be based on pure and incorrigible observation statements that are themselves untainted by theory. With such a scientific language at hand, it should be possible to reduce social sciences to psychology, psychology to biology, biology to chemistry, and chemistry to physics in a unified science. Furthermore, should it be possible to reduce physics to mathematics, and mathematics to formal logic, this unified science would have reached the bedrock foundation of knowledge that was so famously denied by Popper. As they were able to secure academic positions at universities across the United States, the emigrated members of the Berlin and Vienna Circles picked up their old ambitions, publishing the *International Encyclopedia of Unified Science* through the University of Chicago Press, with Otto Neurath, Rudolf Carnap (both former members of the Vienna Circle), and Charles W. Morris from the University of Chicago as its principal editors. Thomas S. Kuhn's "*Structure of Scientific Revolutions*" was first published, in 1962, as a contribution to this "*Encyclopedia of Unified Science*," yet dealing the whole movement a devastating blow. How did this happen?

While still working in Europe, the supporters of the Unity of Science Movement had a liberal social and political agenda. They fled Europe as Germany and Austria drifted to the right and into Hitler fascism. When they found a new home and a new career in the United States, they picked up their old ambitions, which led the movement into a conflict with Cold War ideology in America.[29] With Otto

---

[29]Reisch, G.A. 2007. From "The Life of the Present" to the "Icy Slopes of Logic". Logical Empiricism, the Unity of Science Movement, and the Cold War, pp. 58–87. In: Richardson, A., and T. Uebel (Eds.), The Cambridge Companion to Logical Empiricism. Cambridge University Press, Cambridge, UK.

Neurath's death in 1945, and political pressures from without and within the academic profession increasing, logical empiricism increasingly disengaged from social, political, educational, or humanitarian issues, receding to a technical and neutral attitude in the pursuit of philosophy of science. The work of scientists, and the science that resulted from it, were analyzed in the technical terms that figure in truth-functional semantics, formal logic, and theories of probability. Science and its philosophy were to be a purely academy field of inquiry, barren of all historical, social, and political dimensions.

Thomas Kuhn's "*Structure of Scientific Revolutions*" was a completely unexpected wake-up call for the by then sleeping beauty called "the philosophy of science," dragging her out of the barren desert into which she had receded and bringing her back to life and blood. Kuhn was less interested in the logic that was to be made apparent in Einstein's theories of special and general relativity. He did not believe that the philosophical analysis of science could or should be a matter of axioms, theorems, and laws of logic only. It was, after all, people made of flesh and blood who do the work in science. Kuhn's philosophy of science took its clues from the historical and social context in which science unfolds. Indeed, Kuhn turned to the history and sociology of science to understand its working, its growth, and development. What he found in those historical archives was a picture of science that radically contrasted with Popper's ideas. Some have said that Popper painted a picture of science the way he wanted it to be, in an ideal world, whereas Kuhn canvassed science the way it is, in the real world.[30] Popper caricatured the scientist as a person whose major aspiration was to work toward the falsification of her most cherished theories. Scientists should not attempt to defend moribund theories, but instead adopt a critical, indeed self-critical attitude and look falsification in the eye, discarding old theories when their time was up. He wanted science to grow as a consequence of the replacement of falsified theories by new or modified ones. Although never getting to the ultimate truth about the world, science would nevertheless grow in truth content as a consequence of continued "conjecture and refutation." Einstein's theories might someday be found to be wrong, or incomplete, yet Einstein unquestionably told us more about the working of the universe than Newton had – this was Popper's idea of scientific progress.

Kuhn found nothing of the sort when he sifted through the historical records that trace the working of science. Professors do not generally assign projects to graduate students that seek the mere falsification of the theories they hold in highest regard. Instead, Kuhn found academic teachers to exert a healthy degree of authority over their pupils. Students are brought up in a research tradition deemed by the scientific community of a certain historical period to be worthy of further investigation and refinement. That way, the students mature intellectually while they are embedded in a certain research practice that reflects the social context of the time. Such a

---

[30]Bonde N. 1977. Cladistic classification as applied to vertebrates, pp.741–804. In: Hecht M.K., P.C. Goody, and B.M. Hecht (eds.), Major Patterns in Vertebrate Evolution. Plenum Press, New York, p. 744.

research practice, which defines science in a particular period of its history, partitions the world in ways relevant to the goals of current science. Some philosophers and scientists have read into *"Structure"* the Kuhnian claim that scientists, and through them science, do not simply discover the world as it is, which is the way the logical empiricists and Popper would have it, but rather create this world to some extent through their own activity and theories.[31] Of course, this is not to say that scientists in some way create mountains and rivers, all subject to the Law of Gravity. But it is none-the-less true that professorial authority, steeped as it is in the tradition of a research practice, carves up the world in ways that are relevant, or irrelevant, relative to the prevailing trends in science, and those trends change through time as also does fashion. According to Kuhn, it is the research questions and the scientists that ask them that define what in this world is worthy of further investigation and what is not. Should further research within the traditional scientific practice lead to problems, such as an apparent falsification of a theory or hypothesis, it does not mean that such theories or hypotheses are immediately abandoned and replaced by different ones. Rather, such anomalies are shelved, put aside in the hope that future research will reveal a natural explanation for what at the time of its occurrence is considered a mysterious anomaly. It is only if anomalies accumulate to a degree that seems no longer rationally acceptable that the authority and tradition of a certain special branch of science starts to disintegrate. It is only when the weight of accumulated anomalies becomes too large that science lets go of the old, and gets ready for something new.

Popper concentrated on the logic of the falsification of scientific theories, whereas Kuhn found in the social and cultural history of the scientific enterprise cyclically recurring crises. What he called "normal science" was the fleshing out of a particular branch of science at a particular time in history under a particular paradigm. A "paradigm" is a certain overall world-view, inclusive of a theoretical construct that underlies, directs, and constrains the unfolding of the research of "normal science." It is these paradigms that carve up the world, differently at different times, into domains that are considered to be, or not to be, relevant to current science. According to Kuhn, what the history of science shows is the recurrence of what he called "scientific revolutions." These are times when an old paradigm slides into a crisis as a consequence of too many anomalies having accumulated over time. Such crises trigger "scientific revolutions," times of radical change during which an old paradigm is replaced by the emergence of a new one, complete with a new theoretical foundation that gives science an entirely new direction. Copernicus, Galileo, Newton, Darwin, and Einstein are icons amongst scientific revolutionaries that introduced new paradigms. For Kuhn, scientific

---

[31]The discussions of Kuhn's relativism are numerous, and often left Kuhn with a sense of frustration for being misunderstood. See, for example, Barnes, B., D. Bloor, and J. Henry. 1996. Scientific Knowledge. A Sociological Analysis. The University of Chicago Press, Chicago; Hacking, I. 1983. Representing and Intervening. Introductory Topics in the Philosophy of Natural Science, Cambridge University Press, Cambridge, UK; Scheffler, I. 2002. Science and Subjectivity. Hackett, Indianapolis; Kirk, R., 1999. Relativism and Realism. Routledge, London.

theories are not abandoned because they fail a crucial test, as was the case for Popper. Instead, they fall victim to a scientific revolution that replaces one paradigm, one world-view, with another. The scientist is no longer an ideal investigator free of self-interest, prejudice, and convictions, subjecting theories to the severest tests possible in an attempt to refute them. Rather, scientists are influenced by the authority they experience during their training, and the tradition of a research practice within which their training takes place. The Kuhnian scientist is not a loner who, isolated in her laboratory, goes about the routine testing of pet theories. For Kuhn, science has an irreducible sociological dimension: psychology, sociology, politics all contribute to the shaping of scientists, and to what they do when they go about to play the game called science. This is the juncture at which Miller's fear that the objectivity of science could be compromised by the social and political agendas of scientists gains purchase.

For Popper, the game of science was one of "conjecture and refutation": although there was no truth about the world to be had, science would gain in truth-content, or verisimilitude as he called it, when theories were conjectured, tested, eventually refuted, and replaced by new or modified ones. As Popper would claim somewhat paradoxically, science seeks a growing correspondence of its theories to the physical world, or in other words, scientific theories become increasingly more successful in hooking up with the contents of the world. But this also means that science makes a commitment to the existence of the world it investigates. If scientific theories successfully explain the physical properties of gold, the scientific realist will claim that such atoms with the atomic number 79 actually really exist in the world. Indeed, common sense would seem to support such a realism in science: we trust the art dealer, or jeweler, and expect that the golden objects he/she has for sale are made up of atoms with the atomic number 79, and not of fool's gold. Indeed, people selecting golden wedding bands or ear-rings need not even know anything about the atomic structure of gold, but trust that there are experts in the world, such as chemists and jewelers, who can draw the distinction between real gold and fool's gold. However, here again Kuhn drew a much different picture of what science is all about.

Kuhn's philosophy of science has been criticized for the indeterminacy of his concept of a paradigm and paradigm change. But given that paradigms, and paradigm change, are in part influenced by psychological, social, cultural, and historical factors, such indeterminacy is hard to avoid. The idea, again, is to study science as it unfolds through time, instead of trying to capture a complex historical process in a system of definitions and the logical implications of those. But an equally controversial aspect of Kuhn's philosophy of science is his concept of incommensurability. The concept of incommensurability quite generally is one of lack of communicability: the language of science breaks down as a consequence of a paradigm change. Kuhn thought that theory change during a scientific revolution that results in a paradigm change is so radical that the two generations of scientists – those working before, the others after the revolution – could no longer meaningfully talk to each other. Or, in more general terms, the theories that characterized science before a scientific revolution, and those that are developed during and after a scientific revolution, cannot be exactly translated one into one another. Kuhn considered the theories adopted by pre and postrevolutionary science to be so radically different that the scientists working under those theories would not understand each other anymore. They would talk past one another. Incommensurability thus paints a rather grim picture of science: theories and their meaning appear to be conditioned – at least in part – by the language of science, instead of by the world that scientific theories are supposed to explain. Radical theory change does not (necessarily) mean the replacement of a falsified theory by a new, better one, but the development of a new language of science that expresses a new world-view, embedded in a different social, cultural, and historical context. Science, its method, and its theories would appear to have stronger roots in the sociology and psychology of the scientific community than in the world explored by scientists. According to Kuhn, modern science enjoys no privileged status in the history of science. We may think that modern science got many things right, such that Einstein went beyond Newton in his explanation of the universe. But according to Kuhn, contemporary science could at any time become the subject of a paradigm change, and our best current scientific theories could be replaced by different and incommensurable ones. For Popper, paradoxically for reasons to be discussed in the next chapter, science and its theories approach an understanding of the real world ever more closely as science progresses. For Kuhn, there is no comparable progress of science – there is no real world that exists independent of mind and discourse that science could capture. Scientific theories that explain the world do so always and only in relation to a paradigm, a certain world-view, one that itself creates part of the world that scientists investigate. For scientists, a paradigm works like a pair of tainted glasses. If the paradigm changes, the taint of the glasses changes through which the scientist sees the world, and the world will accordingly look different.

There would be many more influential twentieth century philosophers to discuss, such as Imre Lakatos[32], for example, Popper's successor at the London School of Economics. He sought to combine elements from Popper's and Kuhn's philosophy of science, painting scientific progress as the result of an interplay between competing research programs. Lakatos retained Popper's falsificationist attitude toward scientific theories, but found these embedded in a research program. Its' most basic, general theories form the "hard core" of the research program that should not be abandoned, while the program still proved progressive. Around its "hard core," the research program builds a "protective belt" that consists of testable theories of lesser generality and of relevant observation reports. A research program is judged "progressive" as long as it generates new hypotheses and theories that explain new phenomena. A "degenerative" research program has lost its potential to generate new discoveries. Competing research programs are evaluated in terms of their degree of progressiveness or degeneration. Cliques of scientists who defend competing

---

[32]Lakatos, I. 1970. Falsification and the methodology of scientific research programmes, pp. 91–196. In: Lakatos, I., and A. Musgrave (Eds.), Criticism and the Growth of Knowledge. Cambridge University Press, Cambridge, UK.

research programs provide science with its sociological structure. The reluctance by scientists of letting a research program go in the wake of anomalies generated by tests, and the consequent tendency of scientists to protect its "hard core" from falsification, emphasized Kuhnian values such as authoritarianism and tradition in Lakatos' philosophy of science. But if a research program would start to degenerate for lack of creativity and innovation in the face of a growing body of empirical evidence against it, then it would eventually be replaced by another, new, and progressive research program that asks different questions seeking different answers.

7.6 Paul Feyerabend and Epistemological Anarchism

But certainly the most notorious philosopher of science of the twentieth century must have been Paul Feyerabend[33], if for nothing else than his claim that there is no human activity, clearly distinct from all other social activities, which can be called science. Rumors have it that students enrolling at the University of California in Berkeley, wishing to study philosophy of science under Feyerabend, would be told that there is no philosophy of science, because there also is no such thing as science!

The Austrian born Paul Karl Feyerabend turned to the study of physics, astronomy, and philosophy after World War II, during which he was severely injured. A British Council Scholarship should have allowed him to pursue postdoctoral studies in philosophy with Wittgenstein in Cambridge, but unfortunately the famous philosopher passed away before Feyerabend arrived in England. He consequently turned to Popper as his academic mentor, and while he was initially strongly influenced by him[34], he would eventually become one of Popper's fiercest critics, as he realized that Popperian falsificationism "would wipe out science as we know it."[35] Feyerabend recounts[36] how Imre Lakatos repeatedly encouraged him to review Popper's new book "*Objective Knowledge,*" when it was first published in 1972. Feyerabend declined, as he confessed to have "little interest in Popper's metaphysical excursions." But Lakatos kept insisting, and he needled Feyerabend by sending him all the highly positive reviews of Popper's book that were being published, reviews that triggered in Feyerabend a "slight impulse of retching." When an Oxford philosopher finally compared Popper's style of writing with that of Bernhard Shaw, Feyerabend had enough and sat down to write his scathing critique. It was somehow prevented from being published in the prestigious *British Journal for the Philosophy of Sciences,* allegedly due to machinations which Feyerabend

---

[33]Feyerabend, P. 1995. Killing Time. The Autobiography of Paul Feyerabend. University of Chicago Press, Chicago.

[34]Feyerabend, 1995, ibid., p. 89.

[35]Feyerabend, 1995, ibid., p. 90.

[36]Feyerabend, P. 1981. Probleme des Empirismus. Vieweg, Braunschweig, p. 364.

attributed to John Watkins, "Popper's bulldog"[37] at the London School of Economics (Lakatos and Watkins were in the same department at LSE that Popper was associated with). This episode, taken from Feyerabend, illustrates how intrigue and machinations can influence the nature and course of philosophical debate – and for Feyerabend, the exact same thing was true for science.

Feyerabend characterized the core idea of his philosophy of science as methodological anarchism. Logical empiricists had sought a way to characterize science as distinct from all other social institutions, and Popper thought he had found a methodological answer to the question of what makes science different from metaphysics. According to Popper, science is characterized by its method, and this, the scientific method[38], consists in the conjecture of scientific theories followed by their test in an attempt to falsify them. Again taking his clues from the history and sociology of science, Feyerabend endeavored to show that there is no one single and universally agreed upon method that would characterize the intricate web of intellectual and social interaction that sustains science throughout all its different departments. He consequently titled his first famous book "*Against Method*" (1975), where he branded science with his well-known slogan "anything goes." The German title of the same book is even more telling, as it reads (in translation) "*Against the Enforcement of Methods in Science.*" Science becomes a social construct through and through. Social and political factors determine what science is, and they do so through funding agencies. According to Feyerabend, Voodoo science is not rejected because its theories were found to be false. Instead, societies of Western cultures reject Voodoo science even before having given it a chance to be proven wrong. According to Feyerabend, the game of science is played by power-hungry, egocentric, and greedy scientists seeking fame and fortune. Fame comes in the form of a Nobel Prize, fortune in the form of large research grants that supplement lucrative appointments at prestigious research institutions. Scientists enter into alliances with colleagues and withdraw from those again depending on their interest, needs, and goals. At the political level, for example in panels convened by funding institutions, scientists lobby for the recognition of their own programs as the only truly scientific ones, thus marginalizing alternative attempts to investigate and understand the world and cutting them off from funding. Scientific knowledge is not some sort of objective entity hovering as an aethereal balloon in Popper's imaginary "World of Thoughts and Ideas," ready to be grasped at any time by the investigating scientific mind. Instead, scientific knowledge pretty much reduces to what a clique of scientists claim it to be, most likely in competition with rival cliques of scientists who see the world somewhat differently.

At the end of the day, the study of the ideas of four of the foremost philosophers of science from the second half of the twentieth century leaves us with a rather

---

[37]Musgrave, A. 1999. Orbituary: Professor John Watkins. The Independent (London), August 5, 1999. Watkins was called Popper's "pit bull" by Feyerabend, 1995, ibid., p. 95.

[38]On the "legend" of THE scientific method see Kitcher, P. 1993. The Advancement of Science. Science without Legend, Objectivity without Illusions. University Press, Oxford.

bleak view of science. On the one hand, there is the highly skeptical Popper who denies the possibility of gaining any positive knowledge of the world, or who, as he invites such knowledge into discourse through the backdoor as will be discussed in the next chapter, could do so only on pain of violating his own principles. On the other hand, there is Kuhn and Feyerabend, who characterize science as a social process with all the merits and deficiencies that characterize other social, and political, institutions. If science can be reduced to one of many social institutions, then why should Creation Science, or Intelligent Design theories, not claim scientific status? Naturally, the institutionalized scientific community does not accept "creation science" as proper science, but what is there to stop Creation Scientists of all colors to get organized, and to seek social and political support to get their views eventually accepted as "scientific" – as indeed they try to do? In January 2008, "Creation Science" launched its own journal, the *Answers Research Journal*," with an editorial policy that subjects submitted manuscript to peer-review as is customary for major scientific journals. But of course, according to editor-in-chief, Andrew Snelling, the reviewers will be recruited from a pool of individuals who "support the positions taken by the journal."[39] Such a policy will certainly favor biased, rather than impartial, reviews, which is not to say that bias may not occasionally influence the editorial process of scientific journals also.[40] Feyerabend found the simple dismissal by "established scientists" of Creation Scientists for being "incompetent" or "ignorant" imprudent: "Almost every scientist who puts his findings into a broader context talks about things of which he is ignorant." In support of his claim, Feyerabend offered the story of James Clerk Maxwell, who wrote the entry on the "*atom*" for the ninth edition of Encyclopedia Britannica, published in 1875. In this article, Maxwell conceded that atoms display properties that could only be explained by reference to events that "do not belong to the order of nature under which we live." Indeed, the explanation of those mysterious properties of atoms required a new point of view, a new paradigm – the one that was eventually introduced by quantum mechanics. From this incidence in the history of science, Feyerabend extrapolates to the opinions of evolutionary biologists that clash with the views of proponents of Intelligent Design. The dismissal of Intelligent Design by evolutionary biologists is based, according to Feyerabend, on the belief that biology is already in the possession of knowledge of all the relevant natural laws that are required to explain "the order of nature as we know it." According to Feyerabend's diagnosis, proponents of Intelligent Design do not share this belief: "This is a legitimate difference of opinion that should not be dismissed by relegating it to a no-man's-land called 'unscientific'."[41] As laudable as Feyerabend's humanism and tolerance are that he brings to this debate,

[39]Anon. 2008. Creationists launch "science journal". Nature, 451, p. 382.

[40]For examples see Hull, D.L., 1988. Science as a Process. An Evolutionary Account of the Social and Conceptual Development of Science. The University of Chicago Press, Chicago.

[41]Feyerabend, P. 1982. Votum, pp. 237–238. In: Feyerabend, P., and C. Thomas (Eds.), Wissenschaft und Tradition. Verlag der Fachvereine, Zürich.

something seems intuitively wrong about this view of science. Indeed, while still a student, it was of great importance to Feyerabend to get across the insight that – a belief in God aside – "the idea of a divine Being simply [has] no scientific basis."[42]

The legacy of Popper, Kuhn, and Feyerabend, three philosophers who have been called "irrationalists," with Feyerabend singled out as the "worst enemy of science,"[43] is a deflated view of scientific knowledge. Contemporary science can at best be found to be wrong, and if it is not, it is because psychological, social, and political interests protect it from critical scrutiny. Such a skeptical view of science has led to the famous "pessimistic (meta-)induction from the history of science." Feyerabend, of course, cried out[44] against his newly won title, claiming it was based on a complete misunderstanding of his writing. Kuhn, in turn, deplored a "communication breakdown"[45] between him and his readers, which led him to distinguish two Thomas Kuhns, the real $Kuhn_1$ and the much discussed $Kuhn_2$. Now, philosophical writing often is open to different interpretations, something that is particularly true of Feyerabend's witty and provocative style. But whether intended or not, their legacy is seen by many as precisely this very deflated, indeed pessimistic view of science.[46]

7.7   The Pessimistic View of Science: Reviewed

Quite generally, induction is a method of inference from past experience to the future, which is based on the belief that the future will resemble the past. The claim that the sun will rise again tomorrow is an inductive inference based on past experience of the sun rising every day and the belief that the fundamental structure of the universe, as it was described by Newton, does not change. The picture of science sketched by Popper, Kuhn, Lakatos, and Feyerabend was one of continuous theory change. What past generations considered their best scientific theories have now been recognized as false or incomplete explanations of the world. To judge from the past, there is no reason to believe that our current best scientific theories will fare any better: they, too, will eventually be recognized as false, or incomplete, and be replaced by other ones. In the wake of Popper, Kuhn, and Feyerabend, there

---

[42]Feyerabend, 1995, ibid., p. 68.

[43]Theocharis, T., and M. Psimopoulos. 1987. Where science has gone wrong. Nature, 329: 595–598. Horgan, G. 1991. Profile: Thomas S. Kuhn, reluctant revolutionary. Scientific American, 264: 40, 49. Horgan, G. 1992. Profile: Karl Popper, the intellectual warrior. Scientific American, 267: 38–44. Horgan, G. 1993. Profile: Paul Karl Feyerabend, the worst enemy of science. Scientific American, 268: 36–37.

[44]Feyerabend, 1995, ibid., p. 146.

[45]Kuhn, T.S., 1970. Reflections on my critics, p. 232. In: Lakatos, I., and A. Musgrave (Eds.), Criticism and the Growth of Knowledge. Cambridge University Press, Cambridge, UK.

[46]See, for example, Gauch, H.G. 2003. Scientific Method in Practice. Cambridge University Press, Cambridge, UK, pp. 78ff.

seems nothing fundamentally wrong with the assertion that 90% of the content of our current scientific textbooks will be found to be wrong, and will be replaced by modified or even radically different theories in the future. Now, current textbooks will certainly require revision, and at least partial rewriting – science does not discover eternal truths all at once. But an astronaut would surely hope that the current textbooks on physics and astronomy are correct to a much greater degree than a mere 10% when boarding the space shuttle. And so would probably anybody seeking help in an emergency room for any reason.

The pessimistic (meta-)induction from the history of science is far too pessimistic, too skeptical, it is simply too strong. Should we reject the current version of the theory of evolution, only because it could be found false or incomplete tomorrow? And indeed, could, or would it be found false or incomplete tomorrow? Critics of Thomas Kuhn have countered the pessimistic (meta-)induction from the history of science with the "No Miracle Argument" that goes back to the Harvard philosopher Hilary Putnam. It says that to land people on the moon would have been a miracle if our best current scientific theories were not approximately relevantly true. But miracles do not occur, at least not in the world of science. The "no miracle argument" does not require eternal truth; it is satisfied by scientific theories that are approximately and relevantly true.[47] "Approximately true" because science cannot get to the whole truth all at once, "relevantly true" because scientific theories are of relevance to a certain domain of inquiry, but not to others. The fact that airplanes crash relatively infrequently, and that a crash – should one occur – can be investigated and causally explained, is taken as evidence for the fact that the laws of aerodynamics as known to contemporary science are approximately relevantly true. Future developments in physics may bring about a broadening and refinement of our knowledge of aerodynamics, but it is extremely unlikely that the laws of aerodynamics as we currently know them should at one time be completely rejected or abandoned and replaced by incommensurable new theories. What, then, went wrong in Popper's, Kuhn's, and Feyerabend's, philosophy of science?

The "pessimistic (meta-)induction from the history of science" is still widely propagated in controversial discussions about what is, and what is not, "proper science." Popper finds science to progress toward an ever greater truth-content, or verisimilitude, of its theories, but then claims that theories can never be known to be true; they can only be known to be false as a consequent of their failing a crucial test. Kuhn finds the process of science to cycle through paradigm changes, each paradigm painting a picture of the world that is, at least in part, socially and historically conditioned. Feyerabend reduces science to its sociological and political dimensions, painting the picture of a scientist who pursues fame and fortune in a Machiavellian manner. Kuhn claims that in the course of scientific revolutions, old theoretical constructs are abandoned in favor of new ones, and that such paradigm changes are radical to the extent that old and new theories are incommensurable.

---

[47]Boyd, R., 1991. On the current status of scientific realism, pp. 195–222. In: Boyd, R., Ph Gasper, and J.D. Trout (Eds.), The Philosophy of Science. The MIT Press, Cambridge, MA.

Scientists from before and after a scientific revolution speak a different language, relating to a different world, such that they cannot effectively communicate with one another anymore. How could this be so? When dealing with the pessimistic (meta-)induction from the history of science, some important distinctions have to be drawn. There can be little doubt that science such as physics, engineering, and biology have made enormous progress over the last 200 years: the belief in phlogiston has been replaced by a theory about oxygen; Buffon's theory about "organic molecules" has been replaced by theories of molecular genetics; Newton's world where time and space are different dimensions has been replaced by Einstein's world where time and space form a four-dimensional continuum. But does that mean that Newton's theory was false, and that Einstein's theory is true? Does it mean that if Einstein met Newton in the four-dimensional time–space continuum, they could on such an occasion not converse meaningfully about their respective world-views? Neither alternative seems right. It rather seems that they could easily engage in a discussion of their shared interests, such that Einstein could easily explain to Newton how he built on his theory and expanded it. Einstein did not prove Newton's theory false, but rather added to Newton's theory and thus expanded its scope. Scientists generally do not ask whether a theory is true or false. They rather ask to which degree a theory is confirmed, or disconfirmed by the available evidence. Truth and falsity are concepts of logic that are employed in the philosophy of language. Scientists are generally concerned with probabilities instead.

Popper, Kuhn, Lakatos, and Feyerabend were not practicing scientists, although highly knowledgeable in mathematics and physics, but were philosophers of science. Popper, Kuhn, Lakatos, and Feyerabend did not solve scientific problems, but strove to analyze the way scientists solve scientific problems. Scientists set up experiments and write about those in the language of science. In pursuit of their projects, Popper, Kuhn, Lakatos, and Feyerabend read the books and papers written by scientists in the language of science, and then proceeded to write about those scientific books and papers in the language of philosophy. When analyzing theory change in the history of science, Popper, Kuhn, Lakatos, and Feyerabend took their clues not so much from the language of science, but rather from the philosophy of language. Consider the title of Popper's major book, *"The Logic of Scientific Discovery"*. As its title proclaims, this is a book about "logic." But laws of logic do not govern nature. That is to say, the birds singing in the trees and fish swimming in the sea do not obey laws of logic. Laws of logic govern thought processes as they are expressed in language. Popper says that Einstein's theories have greater truth-content, or verisimilitude, than Newton's, because Einstein's theories explain everything in the world that Newton's theories do, *and more*! But when Popper uses the concepts of "truth," and "falsity" in that context, he uses these terms in a very strong sense (i.e., in the sense logicians use these concepts). Truth in logic is eternal and not tied to time and space. Therefore, what was true yesterday cannot be false today, for if it is false today, it must have been erroneous yesterday. Logicians call such logic "bivalent logic," on which basis a statement, such as a scientific theory, can only have two truth conditions: it is either true or false. From the point of view of bivalent logic, there is nothing in between truth and falsity. A statement is true or false – period! If a crucial test shows a theory to be false today, it cannot have been true yesterday. Therefore, yesterday's theory must have been false. It is the application of such strict bivalent logic to the philosophy of language that results in what is called by its practitioners "truth-functional semantics." And it is because of the fact that Popper, Kuhn, and Feyerabend took their clues from the philosophy of language when they pursued their projects in the philosophy of science that the pessimistic (meta-)induction from the history of science came out far too strong. It is true that Einstein's theory contradicts Newton's in certain special cases. Applying bivalent logic to this situation means that of these two contradictory theories only one can be true, and one must necessarily be false. The conclusion then seems to be that all scientific theories of the past have sooner or later been found to be false in this strict, logical sense, and there is nothing to be gleaned from the history of science that could make us believe that our best contemporary scientific theories will not also be found to be false in the future.

Bringing such strong logic to the philosophy of science paints a black-and-white picture of scientific progress that leaves no room for the gradual growth and refinement of the theories of science. If we relax such strong logical strictures, and side a bit more with the practicing scientists, the inference from the history of science will be far less pessimistic: all previous scientific theories have eventually been found to be incomplete, and in need of revision or amendment. Newton was not wrong – it is just that Einstein pushed the same envelope much further. What Newton once famously said could just as well have been a quote from Einstein: "If I have seen a little further it is by standing on the shoulders of giants." But in order to provide a better vantage point requires the ancestral giants to stand up, not to collapse and vanish.

To understand the growth of empirical sciences that seek to explain the world we live in requires us to relax the strictures of bivalent logic. It requires us to abandon the project of writing a philosophy of science in terms of truth-functional semantics. But even if Kuhn and to an even greater degree Feyerabend were happy to comply with these requirements, the philosophy of language that they employed still holds further surprises in store. What does that mean when we say that a scientific theory is well supported by the available evidence? It means that we compare what the theory says about the world with the world as we experience it. A scientific theory generally picks out some kind of natural objects and explains their properties, dispositions, and their behavior. The scientist attempts to ascertain to which degree a theory is successful in hooking up with the natural object that the theory is about. What a scientific term hooks up with, that is, refers to in the world of experience is called this term's "meaning." And it is a central question for the philosophy of language to investigate how expressions get their meaning. How does a proper name, or a general name, get its meaning? "Oliver Twist," "Mount Everest," or "Alaska" are examples of proper names; "gold," "tiger," "water," or "elm-tree" are examples of general names. Some philosophers of language hold that the meaning of a name is given by the thing(s) the name picks out and introduces into discourse. The name "Bill Clinton" picks out and introduces into discourse the individual baptized Bill Clinton – it is the unique individual that was baptized Bill Clinton that gives the proper name "Bill Clinton," its meaning. But this view is bound to spell trouble. "Santa Claus" is a proper name, "Unicorn" is a general name, and although their use introduces some ideas and mental pictures into discourse, such that people can have discussions about Santa Claus and unicorns, neither name picks out anything in the material world. The sentence "Santa Claus does not exist" is meaningful and evokes disappointment in children, in spite of the fact that "Santa Claus" does not pick out any person in the world of space, time, and matter. So what is it that gives the proper name "Santa Claus" its meaning? Some philosophers of language argue that it is the stories we tell about Santa Claus that gives his name its meaning. For example, Santa Claus is "the bearded man in a red coat that comes through the chimney at Christmas to deliver presents." The same is true for the general name "unicorn." There is nothing in the experienced world that this name could meaningfully pick out and introduce into discourse. Again, it is the stories we tell about unicorns that give the name "unicorn" its meaning, for example, "a horse-like animal with a horn on its forehead."

Kuhn's and Feyerabend's philosophy of science is tied to this so-called "description theory of reference."[48] Take a scientific term like "electron" – how does it get its meaning? Well, we cannot see, smell, touch, or kick around electrons in the way we can kick around rocks. So, the term "electron" is, in some sense, comparable to the term "unicorn": it is the stories that scientists tell about electrons that gives the term "electron" its meaning. In somewhat more technical terms: a scientific term that picks out, that is, refers to entities in the world that we cannot directly observe is called a "theoretical term." According to Kuhn, the meaning of theoretical terms that occur in scientific theories (such as "electron") is determined by the theory itself in which the term is embedded, not by the world the theory attempts to explain. It is the scientific theory that "tells the story" about the properties, dispositions, and behavior of electrons. But now, if a theory changes, then the meaning of the theoretical terms embedded in it changes also. New theories tell different stories about the world than old theories. New theories paint a different picture of the world than the theories they replace. The world thus looks different to those scientists who are willing to adopt the new theory when compared with those elder colleagues who cannot let go from their world view. Scientists who adopt a new theory come to live in a different world, and hence, no longer understand scientists who still adhere to the old theory and vice versa. The scientists talk past each other – incommensurability obtains. For Ancient Greek atomists, atoms were indivisible; the term "atom" itself meant "indivisible." So if we could meet the Ancient Greeks in Einstein's four-dimensional space–time continuum, and if we tried to tell the Ancient atomists about our success in splitting the atom, they would

[48]Devitt, M., and K. Sterelny. 1999. Language and Reality. An Introduction to the Philosophy of Language. The MIT Press, Cambridge, MA. See also Psillos, S. 1999. Scientific Realism. How Science Tracks Truth. Routledge, London, p, 280.

– on Kuhn's account – not understand us. To them, the statement that an atom – "the indivisible" – is divisible would be a contradiction of terms[49], and hence gibberish.

Philosophers of science, especially those with an empiricist background such as Popper, Kuhn, and Feyerabend divided the terms that figure in the language of science into two distinct classes: the observational terms as opposed to the theoretical terms. The division was drawn even if only to criticize it, as was done by Popper. Observational terms are those that refer to, i.e., pick out and introduce into discourse things we can observe in nature. In our earlier examples, "Bill Clinton" is not only a proper name, but also an observational term, since we can observe – at least in principle – the individual Bill Clinton in our world of experience. "Santa Claus" is also a proper name, but it is not an observational term, since we cannot observe Santa Claus in our world of experience (it is for that reason that philosophers call "Santa Claus" an empty proper name). Similarly, "horse" is an observational term, "unicorn" is not. Now let us transpose these simple distinctions into science, say physics or chemistry. "Gold" is an observational term, but "electron" is not. "Billiard ball" is an observational term, but "force" is not. We can pick up or kick away a billiard ball, but we can neither touch "force" or "acceleration" nor touch a "magnetic field." Scientific theories are in some sense nothing but systems of sentences, and the meaning of these sentences (hence also the meaning of scientific theories) is determined by the components of those sentences (i.e., by the words of which these sentences are composed). But the sentences that collectively make up a scientific theory are composed of a mixture of observational and theoretical terms, at least in the eyes of those philosophers, like Kuhn, who want to draw that distinction. So what is it, asks Kuhn, that determines the meanings of the words which figure in scientific theories? Easy to say for observational terms, their meaning is given by whatever they pick out in the world of experience, billiard balls or chunks of gold. This contrasts with "electron" and "force." And here Kuhn says, just like in the case of "Santa Claus" and "unicorn," it is the stories that science tells about electrons, or magnetic fields, that determines the meaning of these terms. Accordingly, the meaning of theoretical terms such as "electron" is determined by the theory in which these terms are embedded, within which these terms function. If the theory changes, then such theoretical terms are either dropped or change their meaning. In either case, the old and new theories are not strictly translatable into one another, and incommensurability obtains.

One example often used to illustrate Kuhn's concept of incommensurability is the phlogiston theory.[50] It is a common observation that things burn only for some period of time, after which the flames succumb and the fire dies. Why is it that some things such as wood burn, whereas other things such as rocks do not? It must be a physical property of wood that renders it combustible. Before oxygen was discovered, it was believed that a special substance called phlogiston was present in those

[49]Rescher, N. 2000. Process Philosophy. University of Pittsburgh Press, Pittsburgh, p. 12.

[50]Kuhn, T.S. 1962. The Structure of Scientific Revolutions, The University of Chicago Press, Chicago, p. 56.

materials that are combustible, whereas the absence of phlogiston would render substances immune to fire. Phlogiston was, of course, invisible, and the name "phlogiston" therefore is treated as a theoretical term. The scientists of the day described phlogiston as a volatile substance inherent in flammable material that was released from the material when it burnt. The flames were thought to succumb when all phlogiston inherent in the burning object was consumed by the fire. However, placing a burning candle under a well-crafted and small enough glass hood caused the flame to die before the candle had burnt to the bottom. And yet, the remaining stump of the candle could be lit again when removed from under the glass. So when placed under glass, the candle's flame died before all the phlogiston in the candle was used up. Phlogiston theorists concluded that air has only a limited capacity to take up the phlogiston that is released by the burning candle. Since there is no new air flowing around a candle placed under glass, the candle stopped burning when the air around it was saturated with phlogiston. This, of course, is the exact opposite of oxygen theory, which says that the candle's flame dies because its burning has used up all the oxygen that was present in the container. So phlogiston theory and oxygen theory tell startlingly different stories about unobservable substances involved with combustion. The meaning of the terms "phlogiston" and "oxygen," if given by the theory in which these terms are embedded, thus turns out to be radically different, indeed contradictory: phlogiston is released, but oxygen is consumed, by the burning candle. When phlogiston-theory was dropped and replaced by oxygen-theory, one language was replaced by another one, one meaning was replaced by a contradictory one, such that incommensurability was bound to obtain between hold-over phlogiston-people and revolutionary oxygen-people.

Kuhn's account of theory change is too radical, the resulting outlook on science too skeptical, all because the description theory of reference to which he ties his philosophy of science is, at least partially, flawed. With his famous *"Naming and Necessity"* of 1972, Saul Kripke radically called into question the description theory of meaning.[51] The book resulted from a transcription of Kripke's Princeton Lectures of 1970, which significantly altered the landscape of philosophy of language, but also earned its authors nasty (yet unwarranted[52]) accusations of plagiarism. Saul Kripke is, indeed, celebrated as a genius amongst living philosophers[53], who published his first technical paper in a professional journal for logic at the age of 19. In 2001, Kripke was awarded the highly prestigious Schock Price in Logic and Philosophy in recognition of his life's work. According to Kripke's philosophy of language, the meaning of scientific terms, both observational and theoretical, is not given by a description that identify the entities picked out by these names, but by the objects themselves. The meaning of "tiger" is the tigers in the world and not the stories we tell about them; the meaning of "electron" is the

---

[51]Kripke, S. 2002 [1972]. Naming and Necessity. Blackwell, London.

[52]Neale, S. 2001. No plagiarism here. The originality of saul Kripke. Times Literary Supplement, February 9, 2001.

[53]Preti, C. 2003. On Kripke. Thomson – Wadsworth, London.

electrons in the world, not the stories science tells about them. It may seem possible to fix the meaning of the term "tiger" by a description of tigers as "large, striped, feline carnivore with four legs and a long tail." In many – indeed most – cases, such a description will correctly identify tigers. However, the problem is that if the meaning of the term "tiger" is invariably fixed by such a description, or, in other words, if such a description were to define the meaning of the term "tiger," then an albino tiger could not be called a tiger even if it was born from tiger parents. And therein lies the crux of the matter: by virtue of being born from tiger parents, the newborn tiger shares a *causal relation* with all other tigers in our experienced world. The tiger cub could be an albino, it could be born with six digits in hands and feet, or it could be born with a limb or the tail missing. Whichever way the tiger cub may have to be described, it is and remains a tiger by virtue of its birth from tiger parents. Or, to put Kripke's analysis more academically, our scientific theories about tigers may change over time, but a tiger is a tiger not in virtue of our scientific theories about tigers, but in virtue of being born from tiger parents.

The description of a tiger as a "large, striped, feline carnivore with four legs and a long tail" is called the stereotype of a tiger.[54] If such a stereotype defined the meaning of the term "tiger," a tiger that freed himself from a trap at the cost of losing one leg could no longer be called a tiger. Worse, remember that supra-natural agent who was to design and create an animal on Mars that looks exactly like an earthling tiger in all respects. That creature would surely fall under a purely descriptive concept of "tiger," even though it shared no causal, in this case no genealogical, relations with any earthling tiger. Scientific theory identifies tigers on the basis of causal relations, such as the relation of descent from tiger parents, and not on the basis of any superficial features. It is therefore an improvement in natural sciences to advance from descriptions to the causal grounding of the meaning of scientific terms. If the concept of "tiger" is causally grounded (i.e., in the common evolutionary origin of the tiger species), then there is no problem in recognizing a three-legged or albino tiger as the descendant from a couple of earthling tigers, whereas the Martian tiger cannot possibly belong to the species of earthling tigers, as it does not share the relevant causal (i.e., evolutionary) relations.

Now, some may say that this is all right and good, but such arguments still do not relieve us from Kuhnian concerns, because "tiger" is an observational term, whereas Kuhn was struggling with theoretical terms such as "phlogiston" or "electron." True enough, but the same argument holds up for theoretical terms. In a science that strives to explain the experienced world, a theoretical term such as "electron" is not taken to be anything but an invention of the human mind, a mere story told in order to make sense of observations a physicist makes in his laboratory, such as those in the classic oil-drop experiment that earned Robert A. Millikan, the 1923 Nobel Prize for identifying the electron as the "carrier" of the basic unit of

---

[54]Putnam, H. 1996. The meaning of 'meaning', pp. 3–52. In: Pessin, A., and S. Goldberg (Eds.), The Twin Earth Chronicles. Twenty Years of Reflection on Hilary Putnam's "The Meaning of 'Meaning'". M.E. Sharpe, Armonk, NY.

negative electrical charge. Instead, scientific realists take the term "electron" to pick out an entity that exists in the experienced world, even if that entity remains elusive to direct observation. And it is because of its elusiveness that the electron cannot simply be described. What can be described are the effects of electrons as they enter into causal relations with each other or with other subatomic particles. Talking about emitters that can spray positrons and electrons, the philosopher Ian Hacking took a nice jab at the description theory of reference and Kuhnian relativism: "if you can spray them, then they are real."[55] "The final arbitrator in philosophy is not how we think but what we do."[56] The meaning of the term "electron" is thus not to be determined by description or scientific theories about electrons, but by the causal relations that electrons are able to enter into. These causal relations are revealed to the experimenting physicists through events in which electrons take part. Our scientific theories about the causal properties, and propensities, of electrons may be wrong, or may change over time, but the electron remains what it is (if it exists at all), and physics will eventually have to adapt its theories to its nature. It is not scientific theories that make an electron what it is through its description; instead, the electron (if it exists) has a nature of a certain kind, and it is the job of physics to discover that nature and to explain what electrons are and do. Physics may approach the nature of electrons, or of any other theoretical entities, in a step-wise manner, such that older theories are revealed to be incomplete and in need of revision or amendments, rather than being dead wrong, or even contradictory to newer theories. Even if not directly observable, physics has no reason to assume that electrons do not exist. Quite to the contrary, earlier theories about atomic and subatomic particles are constantly being revised and refined in the light of new research.

At the end of the day, do electrons exist or not? If you can use them in an experimental context, if you can build an electron gun (it was called PEGGY II[57]) that shoots beams of electrons – how could they not exist? Causality, not description, is the key to reality, and hence the basis for doing natural science. Imaginary entities cannot enter into causal relations. Unicorns cannot attack us. Conversely, even if electrons remain elusive to direct observation, we can assume their real existence if they enter causal relations that can become the subject of scientific investigation and prediction. It is on the basis of the causal dispositions that are typical for electrons that scientific theories about their causal propensities are predictive and consequently testable. This is not the same situation as in the case of phlogiston. Here, the scientific investigation of the phenomena of combustion revealed that phlogiston theory had it the wrong way around: no substance characterized by typical causal propensities can be shown to emanate from all burning objects, phlogiston does not exist, the theory had to be discarded and replaced by a new one. If not for Kuhn, then at least for chemists this represents scientific

---

[55]Hacking, 1983, ibid., p. 23.

[56]Hacking, 1983, ibid., p. 31.

[57]Hacking, 1983, ibid., p. 266.

progress. The terms of chemistry hooked up better with the world of experience once the term "phlogiston" was replaced by the term "oxygen": lack of oxygen will repeatedly, reliably, and predictably extinguish fire.

Present day scientists stand on the shoulders of their predecessors, finding that these predecessors had captured some of the truth about atoms, to which they have been able to add new insights. This is how science progresses. It does not seem to be the case that young scientists do not understand the papers written by previous generations of researchers because of a meaning change of scientific terms that causes incommensurability. It seems rather to be the case that young scientists learn from their predecessors, building on their successes and their failures. Textbooks on physics, or on evolutionary biology, from the 1920s have indeed been revised, and those we have today will likely have to be further revised in the future. But that does not mean that physical theories, or evolutionary theory, as we had them in the 1920s were wrong, and that the physical or evolutionary theories we have today will likewise be found to be wrong in the future. What it means, instead, is that textbooks of physics, or evolution, from the 1920s offered valuable explanations of the natural world, most right, some wrong or incomplete. Today, we have recognized which parts were right, which were wrong, and where the story remained incomplete. And to the parts that were right in 1920s we added new insights, new explanations, thus broadening the scope of physical or evolutionary, theories. Undoubtedly, some of our new, modern insights will again turn out to be wrong, or incomplete, but as can be gleaned from the past, the future is likely to show that most of them were right. And why should that be so? Because scientists are a very competitive lot, some (perhaps even many?) indeed driven by hunger for fame (rather than fortune) – and there is no better way to position oneself in the limelight than by proving one's most illustrious competitor's theories to be flawed in some way or other!

7.8   Description and Explanation: Again

If scientific theories are to provide explanations for the phenomena we experience in the actual world we live in, then they have to be causally grounded. Science is not mere story telling. Science is in the business of investigating and explaining the causal relations that govern natural processes. It is through their causal grounding, through their concern with cause and effect, that scientific theories are predictive (i.e., can make predictions on the future natural course of events). And it is a consequence of their predictiveness that scientific theories are testable. That is not to say that the description of order and regularity in nature is not a scientific achievement, but it is to say that such a description is not also (or already) an explanatory achievement.[58] When we describe the sun as a heavenly body that rises

---

[58]Leplin, J. 1997. A Novel Defense of Scientific Realism. Oxford University Press, Oxford, p. 24.

every morning, we describe a regularity in nature, but in so doing we have not yet explained it causally, as a relation that holds invariably between a cause and its effect(s). To explain something is not merely to describe something; it is to elucidate the causal properties and dispositions of the objects under investigation. With due concern for causal relations, it is possible to understand theory change in science in a far less skeptical, in a far less pessimistic manner than that which was sketched by Kuhn. Most of us are happy with Newton's Universe: we live in a seemingly three-dimensionally organized world that is subject to the passing of time, every fourth year getting an extra day, objects falling down, not upwards, and so on. Yet Einstein was awarded a Nobel Prize for having shown that most of these very basic intuitions are, in fact, wrong. Newton vs. Einstein: a classic case of paradigm change, but not necessarily resulting in incommensurability. True, Newton's mechanics is contradicted by Einstein's theory in certain special cases, but Newton's mechanics still work perfectly well in our everyday world of experience. The picture that emerges from such examples is not one of the wholesale replacements of an old paradigm, or of an old research program, with a new and different one. Instead, it is a picture that shows scientific theories to be complex, multifaceted constructs, where theory change generally results in the partial replacement or refinement of some components of such a construct, whereas other theoretically relevant parts of the construct are retained. The parts of a scientific theory that are retained through a period of theory change are generally those that got the relations of cause and effect that prevail in nature at least approximately and relevantly right. This is how, in the words of the realist philosopher of science Stathis Psillos, "Science tracks Truth."[59]

The answer to the "pessimistic (meta-)induction from the history of science" then is to seek the causal grounding of scientific theory, it is to seek putting causes and effects together. Or, as Wesley C. Salmon put it: it is time to put the cause back into the "because."[60] It has been said that causal relations imply the reality of the objects that enter into these relations. It is the causal grounding of theories of combustion that showed that oxygen exists, phlogiston does not. It is the causal grounding of biological species in evolutionary theory that shows that earthling tigers exist whereas tigers created on Mars do not. Or, to put it the other way around, if it could be shown that tigers created on Mars by a supra-natural agent could produce fertile offspring with tigers on earth, evolutionary theory would be in trouble. Presumably, only existing things can enter causal relations. This means that insofar scientific theories are about the experienced world, and insofar as they are causally grounded, scientific theories tell us what is in the world, or – to speak with the philsopher Quine[61] again – they at least tell us what we must assume to be in the

---

[59]Psillos, S. 1999. Scientific Realism. How Science Tracks Truth. Routledge, London. See also Leplin, 1997, ibid., p. 145.

[60]Salmon, W.C. 1998. Causality and Explanation. Oxford University Press, Oxford, p. 312.

[61]Hylton, P. 2006. W.V. Quine (1908–2000), pp. 181–204. In: Martinich, A.P., and D. Sosa. A Companion to Analytic Philosophy. Blackwell, Malden, MA.

world for our theories to be at least approximately and relevantly true. That causal relations presuppose existing things, and that causally grounded theories make ontological commitments to what there is in the world is easy to understand for material and efficient causes. The material cause is the objects that take part in events, and the efficient cause is the cause that makes events happen. But in the debate between Chambers and Miller, additional causes played a role, the formal cause and the final cause: design, purpose, and goal-directedness.

What, then, is science, and what is scientific? A scientific theory that applies to the world of experience is one that is grounded in material and efficient causes, and that for this reason allows to make testable predictions about the future course of natural events. Scientific theories causally grounded in nature acquire counterfactual force. To take Yuri Balashov and Alex Rosenberg's example: "if the moon were made of plutonium, it would weigh less than 100,000 kg"[62] – the reason being that plutonium spontaneously explodes long before reaching such a mass. The question remains whether such scientific theories could also account for formal and final causes? Natural science answers this question negatively. No "ought to be" can be derived from what "is" in nature. An engineer can ask how an airplane "ought to be" constructed, but he cannot ask how the laws of aerodynamics "ought to be" to make airplane construction not only easy, but also cheap yet still absolutely safe. The laws of aerodynamics just are what they are, and science strives to formulate them as best as possible. If formal and final causes were at work in nature, plant and animal species "ought to be" perfectly adapted; the fact is that they are not. Species are not perfect, but variable instead. The reason why natural sciences in general, and biology in particular, reject final causes is that they presuppose prior knowledge of a goal that is to be achieved by a process. But there cannot be any prior knowledge of any goal for any evolutionary process that would be grounded in any known natural causes. Evolutionary biology also rejects formal causes. Organisms are not built according to a blueprint. Or else that blueprint would have to be massively blurred in order to account for all the variability observed in nature. Formal and final causes do not impart counterfactual force. Cell theory does not allow the assumption that the Creator can work as small as he pleases in support of Bonnet's theory of ovulism, which claimed the encapsulation of preexisting germs of one generation within the eggs of the previous one, since the beginning of time. Conversely, there is no reason to believe that an almighty Creator would at any time suspend the laws of physics and build a celestial body from plutonium that weighs in excess of 100,000 kg.

---

[62]Balashov, Y., and A. Rosenberg. 2002. Philosophy of Science. Contemporary Readings. Routledge, London, p. 42.

Chapter 8
Linking the Facts: Tracing the Traces

Miller based his rejection of theories of species transformation on the fact that no such process of transformation is directly observable, neither in the Fossil Record, nor in living nature. But some of the most exciting explanatory theories of modern science invoke entities that are unobservable: nobody has ever seen an electromagnetic field, or a quark. Such entities reveal themselves through observable effects they have as they participate in causally determined processes or events. Descent, with modification, likewise requires inference from a host of observable traces evolutionary history left behind in fossils as well as in living organisms.

Various methods have been claimed to be characteristic of scientific inference: deduction, induction, and abduction. Deductive inference is the strongest from a logical point of view because it is truth preserving, but is also the least applicable to empirical sciences that investigate an evolving world. Popper's hypothetico-deductivism landed him in an extreme skepticism that he could overcome only at the cost of adopting a hidden element of induction. Inductive inference is weakest from a logical point of view because it is subject to the 'problem of induction', yet it characterizes most empirical sciences that deal with probabilities rather than with necessities. Abduction, also known as 'inference to the best explanation', is close to induction and characterizes not only historical sciences such as evolutionary biology, but also everyday reasoning. It is not necessary to actually have seen the mouse if she left enough traces of her whereabouts to convince us that the time has come to set a mousetrap.

Just as a mouse leaves traces behind on the basement floor, so does descent with modification leave traces behind in the Fossil Record as much as in modern biota. For Darwin, the goal of science was to 'link the facts', to bring as many potentially disparate facts under the same explanatory umbrella. In order to do so, he traced the traces of evolutionary history. The rudimentation of complex organs offered powerful arguments against the doctrine of intelligent design. Biogeography offered powerful arguments for the origin of new species, and insights into how species partition their environment as a consequence of competition. Comparative embryology offered not only powerful arguments in support of a branching order of nature, but also powerful insights into the deep relationships between natural groups of organisms such as reptiles, birds, and mammals. In fact, Darwin found comparative embryology to offer a key to the recognition of common ancestry. A new three-fold parallelism started to emerge between animal classification, the Fossil Record, and embryonic development. Yet not one that runs parallel to the Great Chain of Being, but one that plays out across the branching Tree of Life. Darwin presented a powerful linkage of facts, an intriguing consilience of evidence that rendered the inference to evolutionary explanation a natural one.

O. Rieppel, *Evolutionary Theory and the Creation Controversy,*
DOI 10.1007/978-3-642-14896-5_8, © Springer-Verlag Berlin Heidelberg 2011

8.1   The Problem of the "Uniformity of Nature"

Miller displayed a remarkable stubbornness and bias in his argumentation against theories of species transformation. Descent with modification, he maintained, is not "...standing in experience,"[1] which means that descent with modification cannot be directly observed. As philosophers would say, descent with modification is not directly epistemically accessible. Epistemology is the branch of philosophy that is concerned with the nature and scope of knowledge. For example, my office mate looks out of the window and says: "Better be careful driving home; the streets will be wet tonight." How does he know? He justifies his warning by telling me that as he looked out the window, it was raining, and we both believe that when it rains, the streets get wet. So if it is true that it is now raining, and if it is true that wherever and whenever it rains the streets get wet, then it must also be true that the streets will be wet in an hour from now.[2] If a scientist issues a knowledge claim, his/her peers as much as the broader public will ask him/her to justify the claim, to spell out the reasons he/she has for his/her claim. As we saw in the last chapter, the logical positivists that founded the Vienna Circle, for example, thought that the testability criterion (in their parlance the "verifiability criterion") provides the sufficient reason for making a scientific knowledge claim. For Miller, only direct sensory perception could provide a sound justification for any knowledge claim. Anything that can be an object of direct sensory perception is said to be directly epistemically accessible. But many explanatory theories in various branches of science appeal to entities that are not directly epistemically accessible. Electromagnetic fields, protons, and quarks are not directly observable; their existence is inferred from observable effects such entities have when they interact with other constituents of the world in causally determined processes.

The inference of descent with modification from the observation of nature must be based on arguments grounded in probability or likelihood, but Miller dismissed this as insignificant or at least insufficient. Not yet exposed to the enigmas of quantum mechanics, Miller dismissed probabilistic laws as they did not meet the standards that he required for proper natural sciences. This preceded Herschel's dismissal of Darwin's theory of natural selection as the "law of the higgledy-piggledy." It is the same old argument over again: proper science is built on universal lawfulness because universal laws of nature allow the *deduction* of testable predictions. Given the vagaries of variation and natural selection, natural selection theory does not allow the *deduction* of any specific future evolutionary outcome or event, and hence was dismissed as not properly scientific in nature (however, it does allow probabilistic predictions). Again emphasizing the need to

---

[1]Miller, H. 1849 [1850], Foot-Prints of the Creator: or, the *Asterolepis* of Stromness, Agassiz, L. (Ed.). Gould, Kendall and Lincoln, Boston, p. 278.

[2]The justification is based on *modus ponens*: see Sosa, E. 2000. The raft and the pyramid, pp. 134–153. In: Sosa, E., and E. Kim (eds.). Epistemology, an Anthology. Blackwell, Malden, MA. p. 137.

ground theories in observation, Miller furthermore emphasized: ". . .human observation has not spread over a period sufficiently ample to furnish the required data regarding them,"[3] meaning the spread over time required to observe species transformation. It is in the context of such argumentation that Miller referred to his earlier compatriot, David Hume, who earned himself a most prominent place in the history of philosophy by elaborating on what he recognized as the "problem of induction." The problem here is a logical gap between knowledge based on *past experience* and the predictability of the *future course of natural events*. Is it possible to predict the future course of nature from the accumulated knowledge of the past? Or, to put the same problem the way Karl Popper put it[4]: is it possible to know more than we know?

Traditional empiricism relied heavily on what is known as *inductive* reasoning. What is it, and how does it differ from *deductive* reasoning? Deductive inference starts with a theory, from which testable predictions are deduced. Our standard example has been Popper's: from the universal theory "all ravens are black," the prediction can be deduced "there is no white raven here now" or anywhere. According to the inductivist tradition, science starts with the observation of particular facts, or events. For example, after having pursued bird watching as a hobby for many years, and after having observed many ravens, all of which were uniformly black, one might feel compelled to conclude that "all ravens are black." More generally, if continued observation of particular facts reveals a certain degree of regularity, it seems reasonable to infer that some underlying lawfulness must be the cause of the observed regularity of phenomena. This is Hume's "regularity theory of natural laws," which states that if one particular type of event is constantly followed by another type of event, then actually happening tokens (instantiations, exemplifications) of these two types of events can be assumed to be lawfully linked. Conversely, trust in the underlying lawfulness that is inferred from the observation of past regularity governing the natural course of events would then seem to allow the prediction of future regularity of the same succession of events, such that science would have fulfilled its goal: the generation of successful predictions. Although this argument may seem to be acceptable to common sense, as indeed it seems to work reasonably well in practice, it cannot be backed up in any logically rigorous manner as would seem to be required from a philosophical point of view. From a stringently logical point of view, the scientific endeavor might appear to get entangled in a vicious circle. As Ernst Cassirer[5] pointed out in his discussion of the early development of modern science: how is it possible to infer regularity, by induction, from the observation of particular phenomena, and from there conclude to a future uniformity of law – without getting caught in a circular argument? More technically speaking, Hume recognized the fact that no theory can be *deduced* from

---

[3]Miller, 1850, ibid., p. 278.

[4]Popper, K.R. 1979. Die beiden Grundprobleme der Erkenntnistheorie. J.C.B. Mohr (Paul Siebeck), Tübingen.

[5]Cassirer, E. 1973. Die Philosophie der Aufklärung. J.C.B. Mohr (Paul Siebeck), Tübingen.

observation because, again, deduction is a logical relation that holds between sentences and the thoughts they express, not between words and objects.

In a nutshell, the argument goes as follows: in our daily life, we seem all the time disposed to conclude from past experience to the future course of events, trusting that the natural course of events is, and will remain, uniform. But why should that be the case? Why should the world function the same way tomorrow, or in 1001 years, as it did yesterday, or 100 years ago? To raise this question means to raise the problem of induction. To solve the problem of induction, formulated by David Hume at the philosophical level, would require a logically valid proof that certain generalizations based on past perceptual experiences would also hold in the future. It would require a proof based on *logic* that generalizations such as "the sun always rises in the morning" or "bread is always nourishing for people," based as they are on past experience, will also be true in the future, i.e., that universal laws of nature could be inferred from past regularity. Hume's conclusion was that such a logically valid proof is not available. Appeal is often made in this context to the "Principle of Uniformity in Nature," according to which the natural course of events is governed at all times and everywhere by uniformly acting laws of nature. Such a principle, if valid, would upon superficial inspection seem to justify the claim that causal relations inferred from observed past regularity would allow the prediction of the same regularity for the future course of events. But upon closer inspection the "Principle of Uniformity in Nature" will be recognized as itself being grounded in the claim that experienced past regularity is a reliable guide to the future: the principle already presupposes what it is supposed to explain. This shows the argument from uniformity in nature to simply beg the question of how to justify inductive inference.[6] Yet "begging the question" is considered a classical logical flaw for any argument in which it occurs.

In an earlier chapter, we used a simple syllogism to demonstrate the undeniable power of deduction, which derives from the fact that deduction is truth preserving. If "all humans are mortal," and if "Socrates is human," then "Socrates is mortal." If the premises are true, then the conclusion must necessarily also be true. Alfred Y. Ayer[7] was a British philosopher steeped in the empiricist tradition, who wrestled with the validity of inductive inference. He exposed the fallacy of the "principle of the uniformity in nature" by using it as a premise in such a syllogism, albeit one that concludes from the past to the present. Consider the fact that Europeans had long been acquainted with swans, and all those they had ever seen had been white (as adults). Karl v. Linné, the founder of systematics, named the species of European white swans *Cygnus olor* in 1758. It was the first and only species of swan known at that time, such that it would have been reasonable for someone living in 1758 to conclude inductively that "all swans are white." We have now set the stage to explore Ayer's syllogism that has as its first premise the principle that "(1) nature is uniform everywhere and at all times." As second premise we take the statement that

[6]Ayer, A.J. 1952. Language, Truth, and Logic. Dover, New York, p. 49.

[7]Rogers, B. 2002. A.J. Ayer: A Life. Grove Press, New York.

"(2) all swans observed before the year 1758 are white." If both premises are true, we should be entitled to conclude that "(c) all swans are white." If the first (1) and second premise (2) both are true, then the conclusions (c) that "all swans are white" must necessarily also be true because deduction is truth preserving. However, in 1790, John Latham announced the discovery of black swans in Australia, a species he called *Cygnus atratus*. Evidently, that discovery refuted the conclusion that "(c) all swans are white." This in turn means that one of the premises in our syllogism must be wrong. Yet we *know* that all swans observed before 1758 were white (as adults); the second premise (2) is consequently true. Therefore, the conclusion must be that the first premise (1, "nature is uniform") must be false: the upshot is that there is no deductive justification for the "principle of the uniformity in nature."[8] Hence the insolvable "problem of induction": there is no way to build a logical bridge from the past to the future.

8.2   Popper's Failure to Solve the "Problem of Induction"

Many logical empiricists, Ayer himself included, therefore stopped being concerned about the "problem of induction," at least in its classical Humean formulation, since it revealed itself to be a "pseudoproblem," according to their own standards of what proper science ought to be. In their view, a problem that offers no logically sound solution or any empirically viable test could not be a real, scientific problem, but had to be a pseudoproblem rooted in a misled use of language. This was not the view shared by Sir Karl Popper who, never shy to trumpet his own achievements, opened his book on "*Objective Knowledge,*" published in 1972, with the phrase: "I think I have solved a major philosophical problem: the problem of induction."[9] Popper did so by turning Hume's argument, and with it the world, on its head. Hume had highlighted the problems that obtain as one tries to proceed from observation to theory and on to predictions that would be shown to be true or false by future experience. Popper essentially tried to solve the problem of induction by circumventing it. According to Popper, induction (more precisely, a logic of induction) simply does not exist. It cannot, therefore, play any role in the logical structure of any science. No sound scientific reasoning can possibly be inductive. His famous example is that of the biology professor entering the classroom and telling the students: "observe!" The poor and confused students will naturally have to ask: "But what?" Observation has to start with a question that is generated by theoretical considerations. At the end of his "*Logic of Scientific*

---

[8]The example is taken from Ayer, A.J. 2006. Probability and Evidence. With a New Introduction by Graham Macdonald. Columbia University Press, New York, p. 21.
[9]Popper, K.R. 1973. Objective Knowledge. An Evolutionary Approach, p. 1. University Press, Oxford.

Discovery," Popper emphasized: "Nature does not answer if she is not asked."[10] Students have first to be given a theory to test, before they know what to observe and in which context. Popper starts with posited theories, from which he derived testable observation statements. Recall that what Popper left unexplained is how he would get to the theory in the first place. Thomas S. Kuhn, tongue in cheek, paraphrased Popper's program as one with the goal to "invent theories,"[11] which are then tested against the "real" world, the theory-ladenness of all observation notwithstanding. All that a scientist can do is to invent theories, the bolder the better, and then proceed to test them in an attempt to refute them. A theory that passes the test is not confirmed, only corroborated. This is an important difference! In an footnote added to his "*Logic of Scientific Discovery*" in 1968, Popper emphasized: "I understand the 'degree of corroboration' of a theory as a brief summary of the ways a theory has passed its test, and how severe these test have been. I have never deviated from that position."[12] The degree of corroboration of a theory does not make any promises as to its performance in future tests, for to claim such promises would be a hidden inductive inference: it would mean to conclude from *past experience* (of the performance of a theory under test) to *future experience* (to the performance of the same theory under future tests). As we have seen in the previous chapter, disallowing such inference results in a deep skepticism about science. However, we also noted that Popper claimed science to be progressive, its theories acquiring an increasingly greater "truth-content" or verisimilitude as they are tested, refuted, and replaced or refined, or corroborated. How can this work under Popper's own premises that exclude induction as a logically valid form of inference?

Popper's skepticism is well illustrated by the steadfast Popperian, who claims on the basis of the best scientific theories available at her time that it should not make any difference whether she jumped down the Eiffel Tower expecting to glide safely to the ground with no harm, or whether she took the elevator.[13] How could this be? Well, according to Popper, we cannot know any scientific theory to be true, we can only know it to be false, and the way to find out is to submit the theory to a severe test of a new kind. If a theory fails the test, it will be considered falsified. If the theory passes the test, it will be corroborated. The more severely a theory has been tested, the higher will be its degree of corroboration. But Popper's "degree of corroboration" is strictly only an historical account of how the theory performed in the past, making no promises as to its future performance. Popperian corroboration

[10]Popper, K.R. 1976. Logik der Forschung, 6th ed. J.C.B. Mohr (Paul Siebeck), Tübingen, p. 225; my translation.

[11]Kuhn, T.S., 1974. Logic of discovery or psychology of research, pp. 1–23. In: Lakatos, I., and A. Musgrave (Eds.), Criticism and the Growth of Knowledge. Cambridge University Press, Cambridge, p. 2.

[12]Popper, 1976, ibid., p. 226; my translation

[13]Worrall, J. 1989. Why both Popper and Watkins fail to solve the problem of induction, pp. 257–296. In: D'Agostino, F., and I.C. Jarvie (Eds.), Freedom and Rationality: Essays in Honor of John Watkins. Kluwer Academic Publishing, Dordrecht.

has been compared to an academic transcript, a report that says something only about a student's past performance but that says nothing about the student's future potential. This contrasts with a letter of recommendation that makes promises as to the student's future performance, as does the inductivist's "degree of confirmation" with respect to scientific theories.[14] An inductivist seeks the highest possible degree of confirmation of a theory, such that if a theory was successfully applied in the past, it will be successfully applicable in the future with a high degree of probability. The high degree of confirmation enjoyed by the laws of aerodynamics that allow us to build airplanes today promises us that we will be able to build future airplanes on the basis of the same laws. In contrast, the Popperian believes that progress in science results from the falsification of its most cherished theories, such that new and better theories must be introduced. But if the goal is to falsify a theory, then the expectation must be to devise a test that will possibly bring down a cherished theory. It is true that in the past, all falling objects obeyed Galileo's Law. The degree of corroboration for that law is very high, but that – according to Popper – still says nothing about the future performance of that law in terms of new predictions derived from it. Taking it to its letter, there is nothing in Popper's philosophy that renders it necessary, let alone probable, that Popper's admirer should fall to her death as she jumps off the Eiffel Tower. Or, more precisely, there is nothing in Popper's philosophy that says that taking the elevator is safer, because the elevator is again built on laws of physics that may be found wrong at any time in the future. Now, Popper himself thought that one should base one's action on the most severely tested, most highly corroborated theories of contemporary science, while maintaining that even those could be found false, or at least incomplete, in the future. Rejecting the notion that any scientific theory could ever be known to be true, Popper thought that with a greater degree of corroboration, scientific theories also acquire greater verisimilitude, i.e., they come closer to the truth than theories with a lesser degree of corroboration. In practice, Popper recognized a very high degree of corroboration for Galileo's Law, as well as for the theories of mechanics, which guided the construction of the elevator, and on that basis would have argued that the rational choice to make is to use the elevator rather than to defy Galileo's Law. But as his critics pointed out[15], such recommendation is based on an argument from the past to the future, and thus discloses a hidden element of induction in Popper's philosophy, which Popper claimed to have eliminated from scientific reasoning. However, later in his life, Popper[16] admitted to "a 'whiff' of inductivism" in his reasoning, which "enters. . . with the assumption that science can progress towards greater verisimilitude." W.H. Newton-Smith, one of his critics, had two answers to Popper's admission. First, just as the term "whiff"

---

[14]Godfrey-Smith, P. 2003. Theory and Reality. An Introduction to the Philosophy of Science. The University of Chicago Press, Chicago, p. 68

[15]Putnam, H. 1974. The 'corroboration' of theories, pp. 221–240. In: Schilpp, P.A. (Ed.), The Philosophy of Karl Popper. Open Court, La Salle, IL.

[16]Popper, K.R. 1974. Replies to my Critics, pp. 961–1197. In: Schilpp, P.A. (Ed.), The Philosophy of Karl Popper, vol. 2. Open Court, La Salle, IL; reference is to p. 1193, n.165b.

refers to Atlantic flatfishes, he called Popper's argument fishy. Second, and denying Popper the benefit of charitable interpretation, he called his inductivism not a whiff, but a "full-blown storm."[17]

Fortunately, we live our lives quite successfully, in spite of the capricious arguments proffered by logicians. Whatever logicians tell us about the impossibility to build an infallible bridge from the past to the future, we all build such bridges all the time in everyday life, not infallible ones, though, but quite viable ones. Nobody with good common sense and a cheerful nature who looks forward to a nice *petit apéritif* at the *Café de Flore* on Boulevard Saint-Germain in the early evening hours would choose to jump from the Eiffel Tower instead of taking the elevator. In fact, to make the experience a more memorable one, she might choose to take the stairs down, trusting the weight-bearing capacity of steel. The inference method that we all use in everyday life, but also one that scientists use all the time, and that is based on "linking the facts" has been called "abduction."[18] In fact, there is every reason to call Charles Darwin the "Grand Master of Abduction." In his *"Origin"* of 1859, Darwin paid lip-service to deductive inference, probably to satisfy the standards of scientific theories by physics, and to preempt criticism such as Herschel's. After citing a number of examples of competitive exclusion between species, Darwin concluded: "A corollary of the highest importance may be deduced from the foregoing remarks. . ."[19] But Darwin's argument lacks the rigor and formal structure that is required for valid deductive entailment of a conclusion by its premises. Abduction evidently is neither deduction nor induction – so what is it? Some people call it the "Inference to the Best Explanation." Here is how it works.

8.3   Abductive Inference: The Mouse in the Wainscoting

The example of the *mouse in the wainscoting*[20] was famously articulated by the science philosopher Bas van Fraassen.[21] Remember that Miller, like Hume before him, wanted all science to start from perceptual experience, from observable facts, and found that the theory of descent, with modification, does not live up to such standards. The descent of one species from another one is not directly observable, neither in the Fossil Record, nor anywhere else in nature. But is that such a big

---

[17]Newton-Smith, W.H. 1981 [1994]. The Rationality of Science. Routledge, London, p. 68.

[18]Lipton, P. 2004. Inference to the Best Explanation. Second Edition. Routledge, London.

[19]Darwin, Ch. 1859. On the Origin of Species. John Murray, London, p. 77.

[20]Van Fraassen, B.C. 1980. The Scientific Image. Oxford University Press, Oxford.

[21]The present account is an embellished version of the account given by Psillos, S. 1999. Scientific Realism. How Science Tracks Truth. Routledge, London, pp. 211ff. For a debate see Ladyman, J., I. Douven, L. Horsten, and B. van Fraassen. 1997. A defense of van Fraassen's critique of abductive inference: reply to Psillos. The Philosophical Quarterly, 47: 305–321. Psillos, S. 1977. How not to defend constructive empiricism: a rejoinder. The Philosophical Quarterly, 47: 369–372.

problem for science in general and evolutionary theory in particular? Consider by comparison that there may be reasons bordering on certainty for the assertion that there must be a mouse in the wainscoting, although nobody has actually spotted the mouse (yet). But as his mother would discover, lazy boy Pete did not quite follow his mom's after dinner directions, and left a wrapped piece of cheese on the table, instead of putting it away in the fridge. The next morning, his mother finds the wrapping torn, and an irregularly shaped chunk of cheese carved out from the block. Upon further investigation, she finds droppings on the floor of the kind one would expect to be from a mouse, or at any rate from a small rodent. The droppings point the way to an empty space behind the wainscoting, where she finds newspaper snippings surrounded by such droppings along with empty sunflower seed shells. The seeds seem to have dropped from sunflowers that she had hung from the kitchen ceiling for drying. She concludes that there must be a mouse in the wainscoting, one that seemed to have made the wainscoting its home for some time already. She further infers that the mouse seems to have a particular liking for the kind of cheese from which a piece is missing. She has not observed the mouse, but based on what she does observe, she is inclined to infer the presence of a mouse in the wainscoting. Such inference is the best explanation for all the observations she made, and on that basis she can conclude to the future: in order to get rid of the mouse, she sets up a mousetrap baited with a piece of the cheese that the mouse seemed to like to eat the night before. It seems reasonable to stick with the inference that there had been a mouse in the wainscoting even though the mouse is never caught, in which case it would seem that it had suddenly left for other quarters. On the other hand, the inference that there is a mouse in the wainscoting would turn out to be true if the mouse is found in the trap the next day, but as it is typical for abductive inference, it might as well have been false. Given the observations that need to be linked through explanation, it would certainly seem most likely that it would be a mouse, if anything that would be caught in the trap. But it is not impossible that it could also be the hamster that belonged to the little girl next door, and that had gone missing a few days ago, something Pete's mother had not been appraised of. In the absence of that information, and given the observations she made, Pete's mother was certainly justified to infer that there is a mouse in the wainscoting. By placing a baited trap, she also committed herself to the real existence of the mouse. She did not conclude that the situation merely looked as if there were a mouse in the wainscoting; she thought she could catch and dispose of the real mouse that left real traces of its moving about. In philosophical jargon, the example shows that abduction led Pete's mother to ontologically commit to the existence of the mouse, even though she had not seen it (yet). That is to say: for the inference drawn by Pete's mother to come out true, a real mouse must exist in the wainscoting. Should a hamster turn up in the trap, her inference would still have been approximately true: hamster and mouse are both small rodents. Ontology is the branch of philosophy that deals with issues of existence and being. To make an ontological commitment is to commit to the real existence, in the physical world, of what we are talking about. If abductive reasoning leads to ontological commitments in everyday life, there is no reason to assume that it could not also do so in science.

Remember Henry Gee mentioned in Chapter 5, who in defense of modern methods of inference in paleontological research, and with reference to Popper's scientific method, had claimed that "no science can ever be historical."[22] His claim was challenged by the science philosopher Carol Cleland[23], who in her analysis of experimental vs. historical sciences characterized the first as dealing with repeatable events that can be manipulated in experimental situations, as is indeed required for Popperian tests. Historical sciences, in contrast, deal with "traces" of past events that call out for a causal explanation. The best explanation provided by historical sciences is one that will unify the greatest number of such traces of past events that initially might have appeared to share no common cause. This is just William Whewell's "consilience" again. One of Cleland's examples is continental drift. First proposed by the German geologist Alfred Wegener in 1915, the theory explained the traces left behind by moving continents, such as the approximate match of the western coastline of Africa with the eastern coastline of South America, the similar geology of the southern continental margins on both sides of the Atlantic, and the similar fossils of continental plants and animals (reptiles) preserved in the rocks. Although the theory of continental drift unified all these observations of traces left behind by moving continents, its breakthrough came only in the 1960s, when a causal mechanism (plate tectonics) was proposed that explained *how* continents were (and still are) able to move. The search for the causal mechanism itself was motivated by further traces left behind by earth history, such as the mid-oceanic ridge system, the young geological age of the sea-floor, and the circum-Pacific island arches marking out areas of tectonic activity as revealed by volcanism and earthquakes. When Darwin praised the virtues of "tracing the traces" and "linking the facts," he meant the explanatory unification of similar such traces left behind both by earth history and by descent with modification.

8.4   Darwin: A Master of Abductive Inference

Darwin's 1859 book "*On the Origin of Species*" is a masterful exercise in abductive inference. The philosopher David N. Stamos characterized Darwin's "*Origin*" as "an excellent example of what has come to be known in philosophy of science as *inference to the best explanation*."[24] In his "long argument" against Creationism, Darwin collected a most impressive body of relevant traces of past events from all areas of life history, which he unified in his explanatory account that became known

---

[22]Gee, H. 1999. In Search of Deep Time. Beyond the Fossil Record to a New History of Life. Free Press, New York, p. 8.

[23]Cleland, C.E. 2002. Methodological and epistemic differences between historical science and experimental science. Philosophy of Science, 69: 474–496.

[24]Stamos, D.N. 2007. Darwin and the Nature of Species. SUNY Press, Albany, NY, p. 193; emphasis in the original.

as the theory of evolution. The already mentioned rudimentary hind limbs observed in some snakes are such traces relevant to the theory of descent with modification. "Some of the cases of rudimentary organs are extremely curious," Darwin wrote. "For instance, the presence of teeth in fetal whales, which when grown up have not a tooth in their head... It has even been stated on good authority that rudiments of teeth can be detected in the beaks of certain embryonic birds."[25] How could such observations be explained as perfect adaptation, evidence of design, purpose, and goal directedness in nature? Modern day developmental biologists can take oral epithelium from an embryonic chicken, and bring it together with oral mesenchyme of an embryonic mouse under experimental conditions that will allow normal cell division and differentiation. The surprising result will be that the chicken tissue will engage in the formation of rudimentary teeth.[26] "The eye to this day gives me a shudder...," wrote Darwin to Asa Gray in 1860.[27] The complexity of the eye seemed to defy an explanation of its evolution through variation and natural selection. Instead, its perfection seemed to call for its intelligent design. But, asked Darwin, what about the rudimentation of eyes in "animals inhabiting dark caverns"?[28] Why should an organ designed to perfection succumb to degeneration? Our best current scientific theory on the origin of snakes holds that snakes had a terrestrial origin[29], possibly from a burrowing or secretive lizard-like ancestor. The reasons for the inference that snakes had a burrowing or secretive ancestor are manifold[30], but prominent among these are traces in the anatomy and development of the snake eye that indicate that the snake eye redeveloped from a rudimentary ancestral condition. The snake eye differs in important ways from the eye of a lizard, or from the eye of any other vertebrate animal for that matter[31], which shows that there is more than one way to build a vertebrate eye, whatever its optimal design would be in theory. The cellular composition of the retina is different in lizards and snakes, the lens is different, the mechanisms of accommodation are different, the cells that form the *dilatator* muscle of the iris are recruited from different germ layers in lizards and snakes, and so on. The best explanation for these observations is the hypothesis that the lizard-like ancestor of snakes had a reduced, rudimentary eye, as would be expected in a burrowing or secretive organism, and

---

[25]Darwin, 1859, ibid., p. 450f.

[26]Raff, R.A., and T.C. Kaufman. 1983. Embryos, Genes and Evolution. Macmillan, New York, p. 156.

[27]Darwin, F. 1892. The Autobiography of Charles Darwin and Selected Letters. Dover Edition [1958], New York, p. 220 (footnote).

[28]Darwin, 1859, ibid., p. 454.

[29]Apesteguía, S., and H. Zaher. 2006. A Cretaceous terrestrial snake with robust hindlimbs and a sacrum. Nature, 440: 1037–1040.

[30]Rieppel, O. 1988. A review of the origin of snakes. Evolutionary Biology, 22: 37–130. Conrad, J. 2008. Phylogeny and systematics of Squamata (Reptilia) based on morphology. Bulletin of the American Museum of Natural Historyy, 310: 1–182.

[31]Walls, G.L. 1942. The Vertebrate Eye and its Adaptive Radiation. Hafner, New York.

that living snakes rebuilt their fully functional eye from a rudimentary condition[32].
As a consequence, the snake's eye is not as perfect as it could be: "...The strange
history which their [the snakes'] eyes seem to have had, makes it anything but
presumptive that they have retained the color vision of their lizard ancestors... it is
unlikely that they do, since their cones are plump... and their vision, in consequence, is crude and unsharp as compared with other diurnal vertebrates."[33]

The eye is not the only example of a seemingly perfectly adapted complex
structure. In his fervent defense of evolutionary theory, structured as an argument
against Paley's influential book on "Natural Theology" (published in 1802),
Richard Dawkins[34] used echolocation in bats in this argument that recourse to
Special Creation was not a helpful step to take in the explanation of the evolution of
even highly complex organic systems. Darwin, studying theology at Cambridge
early during his career, was most impressed by Paley's "Natural Theology," and
adopted from this book the notions of design, purpose, goal-directedness, and hence
perfect adaptation, notions that also motivated the writings of Chambers and Miller.
As Darwin later admitted: "I did not at that time trouble myself about Paley's
premises,"[35] and it took him quite some time to break away from this tradition of
thought: "I will here give the vague conclusions to which I have been driven. The
old argument from design in Nature, as given by Paley, which formerly seemed to
me so conclusive, fails, now that the law of natural selection has been discovered.
We can no longer argue that, for instance, the beautiful hinge of a bivalve shell must
have been made by an intelligent being, like the hinge of a door by man."[36] The
hinge of a door was obviously constructed following a preconceived plan and
according to a specific purpose, i.e., to fulfill a specific goal. The same could no
longer be said of a hinge between bivalve shells. Darwin's break with the tradition
of thought then prevalent throughout the establishment of Victorian England was so
radical that his contemporaries seemed simply unable to grasp the full significance
of Darwin's unified explanation of organismic diversity on earth.[37]

Explaining the rudimentation of organs on the basis of purposeful design has
never been very compelling. On February 11, 1793, Carl Heinrich Kielmeyer
(1765–1844), professor at the *Karlsschule* and curator of the natural history collections at Stuttgart, delivered a famous address in honor of the Duke Carl von
Wirtemberg's birthday.[38] Under the title *"On the interrelation of organic forces*

---

[32]Bellairs, A.d'A., and G. Underwood. 1951. The origin of snakes. Biological Reviews, 26:
193–237.

[33]Walls, 1942, ibid., p. 497.

[34]Dawkins, R. 1988. The Blind Watchmaker. Penguin, London.

[35]Darwin, 1892, ibid., p. 19.

[36]Darwin, 1892, ibid., p. 63.

[37]Bowler, P.J. 1988. The Non-Darwinian Revolution. The Johns Hopkins University Press,
Baltimore, MD.

[38]Kielmeyer, C.H., 1793 (1814). Über die Verhältnisse der organischen Kräfte unter einander in
der Reihe der verchiedenen Organismen, die Gesetze und Folgen dieser Verhältnisse. Christian
Friedrich Osiander, Tübingen.

within the Great Chain of Being, and the laws and effects of these interrelation," he described in a most attractive manner the law of harmony determining all processes of life, a "Law of Compensation," which requires that an increase in any one component necessitates a decrease in some other force. Looking back on the Aristotelian world of dynamic permanence once again, he emphasized that although organisms may seem to change throughout their life, this change is only apparent. Instead, the organism persists in a harmonious equilibrium, subject to the dynamic permanence of checks and balances. The human observer may observe changes, but this is an illusion, created by the material appearance of living beings; in essence, organisms remain essentially one and the same throughout their lives, a system in equilibrium. It so happens that Georges Cuvier was one of the many prominent students at the *Karlsschule* in Stuttgart, and it may well be that the Aristotelian idea of a dynamic permanence, translated by Kielmeyer into the language of contemporary life sciences, influenced Cuvier in his life-long campaign against species transformation.[39] Cuvier was brought to the Paris Museum by Geoffroy Saint-Hilaire,[40] who viewed Kielmeyer's Law of Compensation from an expanded angle. While individual organisms were subject to dynamic permanence, Kielmeyer nonetheless hypothesized that forces active in development could, by analogy, provide a material explanation for the parallelism of embryonic development and the Great Chain of Being. He further recognized that different forms of organization, aligned along the *scala naturae*, came into being at different times in earth history, and considered the possibility that the same force which guided the successive appearance of forms of organization would also have been responsible for the origin of life on earth.[41] For Etienne-Geoffroy Saint-Hilaire, Kielmeyer's "law of compensation" would range over the transformation of species, but within strict limits. Geoffroy spoke of a "*Loi du Balancement des Organes,*"[42] a law that says that if one part of an organism increases in size or number, such increase must be compensated by the decrease in number or size of other parts of the same organisms. Compare lizards with snakes: the loss of limbs is compensated by an increase in the number of vertebrae, resulting in an increased body length. This is Hugh Miller's example for his notion of "homological symmetry of organization,"[43] which is nothing but Geoffroy's "Law of the Balance of Organs" recast as a consequence of purposeful and goal-directed design. Taking up these ideas, Darwin noted "In works on natural history rudimentary organs are generally said to have been created 'for the sake of symmetry,' or in order to

---

[39]Lefèvre, W. 1984. Die Entstehung der biologischen Evolutionstheorie. Ullstein, Wien, p. 104.

[40]Appel, T.A. 1987. The Cuvier-Geoffroy Debate. French Biology in the Decades before Darwin. Oxford University Press, Oxford.

[41]Kielmeyer, 1793 (1814), ibid., p. 41. For an attribution of a theory of species transformation to Kielmeyer see Balss, H. 1930. Kielmeyer als Biologe. Sudhoffs Archiv für die Geschcihte der Medizin, 23: 268–288.

[42]Geoffroy Saint-Hilaire, E. 1830. Principes de Philosophie Zoologique, discutés en Mars 1830, au sein de l'Académie Royale des Sciences. Pichon & Didier, Paris, p. 215.

[43]Miller, 1850, ibid., p. 180; the law of compensation is also implied on page 119.

'complete the scheme of nature'[44]; but this seems to be no explanation, merely a restatement of the fact."[45] In other words, Darwin found such explanations to be logically flawed because they are question begging. The explanations already presume the facts they are meant to explain. "Would it be thought sufficient to say that because planets revolve in elliptic courses round the sun, satellites follow the same course round the planets, for the sake of symmetry, and to complete the scheme of nature?"[46]

Darwin collected the traces, the keys to the theory of descent, with modification, from a vast variety of sources: domestication, the empirical study of variation, plant and animal breeding, animal behavior, paleontology, comparative anatomy, and comparative embryology. Yet a particularly important source of relevant traces begging for an explanation was biogeography, the geographical distribution of plants and animals. "Geographical Distribution" is the only title that deserved two chapters in Darwin's "*Origin*" from 1859. It is well known how Darwin agonized over the development of his theory, how he held back from publication for years, assembling more and more evidence in its support. He was finally pushed to rush an abbreviated account of his findings to press when alerted to the fact that another naturalist had independently discovered the principle of natural selection. How could that have happened? Alfred Russell Wallace was predominantly concerned with issues of biogeography, and it is those that pushed him toward a theory of natural selection. When the two men realized that they had been working on the same theory, drawing similar conclusions, they agreed for a paper from each author to be read at a meeting of the Linnean Society of London on July 1, 1858.[47] At that time, Darwin was hastily putting together his "*Origin of Species*" to be published the following year. Working in the Indoaustralian archipelago, Wallace would make a discovery, which today bears his name and which, indeed, beautifully supported the idea of species transformation.[48] What he discovered was the so-called "Wallace Line," a seemingly invisible and untouchable boundary separating the faunas from the small islands of Bali and Lombok from one other. This boundary was represented by a seaway between these two islands of only about 35-km width. The fauna northwest of that line shows distinct similarities to the fauna of the East Asian mainland, whereas southeast of the "Wallace Line" faunal elements show closer affinities to Australian elements. What Wallace, in fact, had to deal with were neighboring islands of the Indoaustralian archipelago, situated next to each other and hence very similar in terms of their physical environment, but differing fundamentally in their faunal composition. This seemed to clash with a

---

[44]The remark "to complete the scheme of nature" refers to Agassiz' interpretation (in his "*Essay of Classification*" from 1857) of limb-reduced lizards as merely rounding out the "type" of reptiles for reasons of continuity and plenitude in the Creation. See discussion in chapter six.

[45]Darwin, 1859, ibid., p. 453.

[46]Darwin, 1859, ibid., p. 453.

[47]Spearman, R.C.I., 1988. Two hundred years of scientific meetings. The Linnean, 4: 30–32.

[48]Winsor, M.P.1991. Reading the Shape of Nature. Comparative Zoology and the Agassiz Museum. The University of Chicago Press, Chicago, pp. 250f.

natural law he had published earlier, namely that *"every species has come into existence coincident both in space and time with a pre-existing, closely allied species."*[49] Support for this principle Wallace drew from the fact that "most closely allied species are found in the same locality or in closely adjoining localities, and. . . therefore the natural sequence of the species by affinity is also geographical."[50] But why, then, this faunal discrepancy on these two closely juxtaposed islands? The only explanation possible for these observations would seem to be that the faunal elements north of the "Wallace Line" had historical affinities to the biota of India, and those south of the "Wallace Line" shared historical affinities with the biota of Australia. There had to have been the opportunity for animals to migrate south from India and north from Australia across this island arch, and in the process they would have evolved into different species in adaptation to different environments. The two faunas with different historical origin apparently met at the "Wallace Line," where they seem to have been kept separate during the geological past by a persisting barrier, a deep sea channel.[51]

Darwin reached similar conclusions when pondering what appeared to be competitive exclusion between species, ". . .namely, that the structure of every organic being is related, in the most essential yet often hidden manner, to that of all other organic beings, with which it comes into competition for food or residence, or from which it has to escape, or on which it preys."[52] After having sought confirmation from the expert ornithologist John Gould of his initial intuition that occurred to him over dinner at a camp fire during the Beagle voyage, he reported in his *"Origin"*: "The plains near the Straits of Magellan are inhabited by one species of Rhea (American ostrich), and northward the plains of La Plata by another species of the same genus; and not by a true ostrich or emu, like those found in Africa and Australia under the same latitude"[53], i.e., in similar environmental conditions. Why would that be so? Evidently, the true African ostrich and the Rhea must have a more distant evolutionary relationship as opposed to the northern and southern species of American ostriches (both in the genus *Rhea*), which would share a much closer evolutionary affinity, yet remain separated from one another in their geographical distribution due to competitive exclusion.

Darwin's visit to the Galapagos Islands is legendary. As his fellow scientists back in England would point out once they had completed the study of the material Darwin had brought back from the Beagle voyage, the Galapagos Islands harbored many traces of species transformation. Located not too far away from Ecuador yet of volcanic origin, the islands harbored species of turtles and finches that appeared to be closely related to continental species, yet had undergone quite striking

---

[49]Wallace, A.F. 1855. On the law which has regulated the introduction of new species. Annals and Magazine of Natural History, (2) 16, p. 186.

[50]Wallace, 1855, ibid., p. 185.

[51]See Winsor, 1991, p. 251f, for more details.

[52]Darwin, 1859, ibid., p. 77.

[53]Darwin, 1859, ibid., p. 349.

adaptive modifications. Darwin collected a sample of finches on the Galapagos Islands and turned them over to the ornithologist John Gould from the Zoological Society of London for systematic description.[54] He commented on Gould's results in his account of the Beagle voyage: "...if Mr. Gould is right [in his systematic conclusions]... there are no less than six species with insensibly graduated beaks."[55] The conclusion seemed obvious. The finches, perhaps only one species, seemed to have invaded the archipelago coming from the South American mainland.[56] On the Galapagos Islands, they found opportunities for radiation into various ecological niches. Since no species had a beak absolutely identical to that of its neighbor, and since population pressure promoted diversification in the use of food resources, natural selection would work on beak variation and eventually cause divergent trends in beak size and shape between populations. This model of species origination seemed supported by Gould's conclusion as to what the species and their relationships are. Today, the finches on the Galapagos Islands, now called "Darwin's finches," provide one of the most important and powerful systems for the study of evolution at the species level. Darwin summarized his findings by noting that "If we look to the islands off the American shore, however much they differ in geological structure, the inhabitants though they may be all particular species, are essentially American... We see in these facts some deep organic bond, prevailing throughout space and time, over the same areas of land and water, and independent of their physical conditions. The naturalist must feel little curiosity, who is not led to inquire what this bond is. This bond, on my theory, is simply inheritance..."[57]

8.5 Darwin and the Importance of Embryology

The human body itself reveals traces of its historical past[58]: why should it be that the human embryo forms a notochord, when in the adult it is replaced by a vertebral column? Darwin thought that embryology "...rises greatly in interest, when we thus look at the embryo as a picture, more or less obscured, of the common parent form of each great class of animals."[59] As mentioned earlier, he found embryos of toothless whales to develop vestigial teeth. Yet, these teeth never become functional. So why should they differentiate in the first place, what purpose, and which

[54]Bowler, P.J., 1984. Evolution, the History of an Idea. The University of California Press, Berkeley, p. 153.

[55]Darwin, Ch. 1962. The Voyage of the Beagle. The Natural History Library Edition, Doubleday & Co. Inc., New York, p. 380.

[56]Sato, A., H. Tichy, C. O'hUigin, P.R. Grant, B.R. Grant, and J. Klein. 2001. On the origin of Darwin's finches. Molecular Biology and Evolution, 18: 299–311.

[57]Darwin, 1859, ibid., p. 349f.

[58]Shubin, N. 2009. Your Inner Fish: A Journey into the 3.5-Billion-Year History of the Human Body. The University of Chicago Press, Chicago.

[59]Darwin, 1859, ibid., p. 412.

goal could they serve? A human embryo develops on either side of its pharyngeal region four outward pouches called "arches," separated by furrows. At this stage of its development, it resembles with respect to those structures the initial stages of gill development in fishes. In humans, however, gills never develop, but a lung instead. Agassiz had spoken of a lawful parallelism of embryological development, Fossil Record and classification, but this three-fold parallelism merely stated a pattern of order in nature, it only described what Agassiz believed to have observed. To prefix the concept of a "three-folded parallelism" with the notion of "Law" does not do any explanatory work, and it does not link any facts as true explanations link causes to effects. It, again, is a question begging argument, and again it was Darwin who could score the point for his theory, as he provided a causal explanation of the observed phenomena: similarities of early developmental stages were to be explained by common ancestry – differences of later developmental stages were the result of descent with modification.

Karl Ernst von Baer, the leading early nineteenth century embryologist (familiar from chapter four), had admitted in his 1828 memoir *"Entwickelungsgeschichte der Thiere"* ("On the Development of Animals") that he had stored early embryos of lizards and birds in unlabeled flasks. Going back to his collection of embryos at a later time, he found himself unable to identify these embryos even to the class to which they belong: which one was the bird and which one the reptile?[60] Von Baer took this as evidence for hierarchical order in nature: lizards and birds each form a group, both of which belong to a more inclusive, i.e., higher group of organization that shares a certain type of construction as is reflected in the similarity of their embryos. With that statement, von Baer offered a most important insight, but again, that insight was nothing but descriptive in nature. Darwin went one step further: he *explained* hierarchical order in nature by common descent. In fact, Darwin – who initially attributed the mix-up of embryos to Louis Agassiz[61] – took this as the best proof of the fact that "embryos... of distinct animals within the same class are often strikingly similar." Why should that be so? Well, because of common ancestry. Why was it the case that "certain organs in the individual, which when mature become widely different and serve for different purposes, are in the embryo exactly alike?"[62] Common descent and subsequent modification is the natural explanation. Once again, von Baer had taken an important step toward theory construction in biology. He recognized the hierarchical structure of development, i.e., a certain pattern of order in nature. Chambers, in his *"Vestiges,"*[63] was the first to translate

---

[60]Baer, K.E.v. 1828. Über Entwickelungsgeschichte der Thiere. Beobachtung und Reflexionn. i. Theil. Gebrüder Bornträger, Königsberg, p. 221.

[61]Darwin, 1859, ibid., p. 439. For the history of this puzzling error see Oppenheimer, J. 1968. Embryological enigma in the Origin of Species, pp. 292–322. In: Glass, B., O. Temkin, and W.L. Strauss jr. (Eds.), Forerunners of Darwin, 1745–1859. Johns Hopkins University Press, Baltimore.

[62]Darwin, 1859, ibid., p. 439.

[63]Secord, J.A. (Ed.) 1994. Robert Chambers: Vestiges of the Natural History of Creation, and Other Evolutionary Writings. The University of Chicago Press, Chicago, p. 212.

von Baer's argument into a simple branching diagram.[64] But to recognize a pattern is not yet to explain it, and important scientific theories are always explanatory theories. It is rare in biology that an author publishes a two volume memoir, which would not only earn him a prominent place in the history of science but also would secure him a prominent role in modern discussions of evolutionary principles almost 200 years after his work appeared in print. Von Baer achieved this goal, although he remained a staunch opponent to Darwin's evolutionary theory till the end of his life. Ironically, it was his fate that in his defense of purpose and goal-directedness in nature, he provided some of the most powerful arguments in favor of theories of species transformation.

With his embryological investigations, von Baer set out to debunk the "Meckel – Serres Law," according to which the embryo recapitulated during its development the Great Chain of Being, the ascending ladder of life that runs from mushroom to human and that so much intrigued Chambers and Miller. Along with Cuvier, von Baer was perhaps the one pre-Darwinian author who contributed most to the replacement of the picture of a ladder by the picture of a branching hierarchy in the depiction of the order of nature, but again, both only pictured that hierarchy, they did not causally explain it as a Tree of Life that had grown through geological time. Logic can be used to dichotomize the world into tables and chairs, cats and dogs. Both Cuvier as well as von Baer found logic, that is, dichotomous hierarchical order, to permeate nature. Amongst animals with some sort of a backbone, however primitive, a fist division separates jawless vertebrates from jawed vertebrates. A second division separates cartilaginous fishes from bony fishes. A division within bony fishes separates lobed-finned fishes from land-dwelling tetrapods. Further divisions separate amphibians from amniotes, reptiles and birds (sauropsidans) from mammals, and great apes from humans. Von Baer found the hierarchy of the animal kingdom that had been worked out by Georges Cuvier and his colla-borators reflected in the embryonic development of animals. Once again we find in von Baer's writing the "type of organization" as an abstract concept: "The devel-opment of the embryo relates to the type of organization as if it [the embryo] passed through the animal kingdom according to the *méthode analytique* [logical method] of the French systematists,"[65] of which Cuvier was the most prominent exponent. But if the natural order is hierarchical, if the animal kingdom and embryonic development are hierarchically ordered and structured, then nature could be ordered not along a linear series of increasing complexity as formalized by the Great Chain of Being, but had to be ordered in a hierarchically branching pattern. To leave the Great Chain of Being behind, and to advocate a hierarchically structured branching pattern both for embryonic development and classification is von Baer's lasting contribution to evolutionary biology. The importance that the recognition of a

---

[64]Richards, R.J. 1992. The Meaning of Evolution. The Morphological Construction and Ideologi-cal Reconstruction of Darwins Theory. The University of Chicago Press, Chicago.

[65]Von Baer, K.E. 1828. Entwickelungsgeschichte der Thiere, vol. 1. Gebr. Botnträger, Königsberg, p. 225.

branching pattern had for Darwin's thinking cannot be overstated.[66] In his 1837 notebook on species transmutation, Darwin famously for the first time sketched a branching diagram, i.e., a family tree or, more precisely, a phylogenetic tree where lineages split as one ancestral species gives rise to two (or more) descendant species. Next to that crude little diagram he jotted down: "I think." The theory of species transformation is therefore no longer just a theory of the transformation of one species into another along a linear gradient of increasing complexity but the splitting and hence multiplication of species through time. That is how biodiversity increased through time, and that is how to explain the fact that species represented by fossils do not only change in successive layers of sedimentary rock, but also increase in number and diversity.

Von Baer found the logic of the order that in his view permeates nature to reflect thoughtfulness and a purpose that was fulfilled by the goal-directedness of embryonic development. To capture the order of embryonic development, von Baer formulated four laws of development, two principal and two auxiliary ones. The two principal ones state: "1st: The general features of a large group of animals appear earlier in the embryo than the special features; and 2nd: Less general characters are developed from the most general, and so forth, until finally the most specialized appear."[67] What these two laws are saying is that, for example, all vertebrates share a structural plan that is laid down very early in development. All the early vertebrate embryos (or at least those that will develop jaws, i.e., gnathostome embryos) show a division of the body into a head, a trunk, and a tail; they show the pharyngeal pouches discussed earlier; they show four limb buds arranged in two pairs; they have primordial vertebrae that are called somites, arranged along the notochord, an elastic structure that underlies the spinal cord, etc. From this common body plan, the fishes are the first to deviate during subsequent development, whereas the embryos of reptiles, birds, and mammals continue to share similarities into later developmental stages. For example, from the limb buds develop various types of fins among fishes, whereas the early developmental stages of the pentadactyle limb are remarkably similar in amphibians, reptiles, birds, and mammals, the latter referred to as tetrapods.[68]

8.6 Clinching the Argument for Evolution

One of the most famous, early, and therefore classic embryological studies based on von Baer's laws, and a good illustration of their meaning, is that of Karl Bogislaus

---

[66]Ospovat, D. 1981. The Development of Darwin's Theory. Natural History, Natural Theology, and Natural Selection, 1838–1859. University of Cambridge Press, Cambridge, UK.

[67]Translation from Gould, S.J. 1977. Ontogeny and Phylogeny, Harvard University Press, Cambridge, MA, p. 56.

[68]Shubin, N.H., and P. Alberch. 1986. A morphogenetic approach to the origin and basic organization of the tetrapod limb. Evolutionary Biology, 20: 319–387.

Reichert, dating back to 1837.[69] To render Reichert's technical account more accessible, and to place it into its broader context, it will here be paraphrased and fleshed out a little. Once again we return to those notorious pouches, the arches that grace the pharyngeal region of sharks and humans during early stages of their development. In fishes, these structures will give rise to gills, which are supported by skeletal structures, the gill arches. The gill arches are first laid down in cartilage, but may later ossify. Each gill arch has a number of components, of which the most important ones are an upper (dorsal) and a lower (ventral) part. Reichert was interested in the developmental fate of those gill arches, in particular of the first two, since these showed important modifications during later development. Looking at generalized jawed vertebrates, such as sharks, Reichert found that the first two gill arches would never support gills, or only in part in the case of the second arch. For these reasons, he named the first two arches "visceral arches," to distinguish them from the true gill arches, the "branchial arches." He further found that in the shark, the upper part of the first (visceral) arch would form the upper jaw, and the lower part of the same arch would form the lower jaw. The upper and lower components of the second (visceral) arch would then attach to the jaw joint, such that the upper part of the second arch would play an important role in the suspension of the jaws from the braincase. So the upper part of the second arch does not serve any hearing function, but instead is an important structural link in the suspension of the jaw apparatus from the skull. The same is true for all fishes.

    True to von Baer's laws, Reichert found lizards, pigeons, and mice to also show the presence of two anterior arches, i.e., the two visceral arches, during early developmental stages. However here, important changes, called "metamorphoses" by Reichert, occur during later stages in development. Most importantly, the upper part of the second (visceral) arch no longer functions as a structural link in the suspension of the jaw apparatus from the skull. Instead, it develops to form a slender ear ossicle, the stapes, which serves the transmission of airborne sound from the tympanic membrane to the inner ear organ. This is concordant with the fact that reptiles, birds, and mammals are all included in a natural group called amniotes that does not include sharks or any other fishes.[70] While the upper part of the second visceral arch transforms to a sound-transmitting element in the lizard, pigeon, and mouse, the jaw joint still remains located between remnants of the upper and lower parts of the first visceral arch in reptiles and birds. This is concordant with the fact that reptiles and birds form a natural group, the sauropsids, which does not also include the mammals. The most surprising observations made by Reichert revealed that in the mouse, both the upper and lower components of the first (visceral) arch loose all connection to the jaw apparatus during later stages of development, and instead develop to form ear ossicles. Amongst all vertebrates, mammals are unique

---

[69]Reichert, C. 1837. Über die Visceralbogen der Wirbelthiere im allgemeinen und deren Metamorphose bei den Vögeln und den Säugethieren. Archiv für Anatomie, Physiologie, und wissenschaftliche Medizin 1837: 120–222.
[70]For the sake of simplicity, amphibians are left out of this account.

in that they have three ear ossicles. The innermost is the stirrup (stapes), which develops from the upper part of the second visceral arch. The intermediate one is the anvil (incus), which develops from the upper part of the first (visceral) arch. The outermost ear ossicle is the hammer (malleus), which develops from the lower part of the first (visceral) arch.

To von Baer and Reichert, the concordant hierarchies of classification and development indicated "degrees of affinities": reptiles, birds, and mammals share closer affinities with each other than either shares with sharks or any other fishes. Reptiles and birds share closer affinities than either of them share with mammals. But these affinities were not considered to reflect different degrees of evolutionary relationships as depicted by a branching diagram. These affinities were thought to be abstract concepts reflecting a thoughtfully designed and logically structured hierarchical order. And yet, this is the point at which von Baer, and Reichert, can be credited with a major scientific achievement, namely the discovery of a theoretically highly relevant pattern in nature. Where they failed, however, is in the search for a causal explanation of this pattern. Reichert recognized the theoretically relevant correlation of parts in the skull of a mouse and a shark, where embryonic development showed the outer ear ossicle of the mouse (hammer, malleus) to correspond to the lower jaw of the shark, the middle ear ossicle of the mouse (anvil, incus) to correspond to the upper jaw of the shark, but he concluded his study with the mere statement of that correspondence without seeking a causal explanation for it. Once again it was Darwin who filled in the missing pieces, thus earning well-deserved recognition for a unifying explanatory theory: "Descent being on my view the hidden bond of connexion which naturalists have been seeking under the term of the natural system. On this view we can understand how it is that. . . the structure of the embryo is even more important for classification than that of the adult. For the embryo is the animal in its less modified state; and in so far it reveals the structure of its progenitor. . . Thus, community in the embryonic structure reveals community of descent."[71]

Darwin's explanation of the traces of past history in the embryonic development of organisms was beautifully vindicated by subsequent research on the origin of mammals. Consider the fact that the jaw joint of sharks, as of all vertebrates with jaws except for mammals, is located between the upper and lower components of the first (visceral) arch or their remnants, both first laid down in cartilage. According to Reichert's observations published in 1837, this primary jaw joint has been transposed into the middle ear in mammals: it is now the point of articulation between the hammer (malleus) and the anvil (incus), both again first laid down in cartilage. But this means that mammals had to evolve a new, secondary, lower jaw joint. They did so by using two different bones, both not preformed in cartilage, but ossifying directly in deep layers of the skin. The bones in question are the squamosal in the skull, on which articulates the dentary that forms the lower jaw. Among all vertebrates, only mammals have a secondary jaw joint formed by the squamosal

---

[71]Darwin, 1859, ibid., p. 449.

and dentary. However, according to Darwin's theory, this structure had to evolve in a gradual, step-wise manner, as is required by variation and natural selection, which was "daily and hourly scrutinizing throughout the world, every variation, even the slightest; rejecting what is bad, preserving and adding up all that is good; silently and insensibly working, whenever and wherever opportunity offers, at the improvement of each organic being..."[72] It so happens that since the days of Darwin, paleontology has delivered an entire sequence of Permian to Lower Jurassic fossils that perfectly document every step of the evolution of the mammalian secondary jaw joint. Some of these fossils are quite striking, such as *Diarthrognathus* from the Lower Jurassic (approximately 200 million years before the present) of South Africa that was first described in 1958 by A.W. Crompton[73], then at the South African Museum in Cape Town, later a renowned professor of paleontology at Harvard. In the name *Diarthrognathus*, the syllable *"Di-"* alludes to a duplicated structure, *"arthro-"* alludes to "joint," and *"gnathus"* alludes to "jaw." The fossil is so named because it has both jaw joints simultaneously, the one formed by remnants of the first (visceral) arch lying immediately deep to the new joint formed by the squamosal and dentary. And as if that was not enough to vindicate Darwin, F. H. Edgeworth[74] discovered in marsupial offspring that during the first 3 weeks after their early birth at a somewhat immature stage, the lower jaw is supported by elements that form the primary jaw joint (i.e., components of the first [visceral] arch) in non-mammalian vertebrates.[75] The secondary jaw joint is not yet formed at that stage. This means that in marsupials, the shift from the primary to the secondary jaw joint occurs at a postembryonic and hence functional stage of development. With the work of von Baer, Darwin, and later authors, a new threefold parallelism emerged, yet one that plays out across the branching Tree of Life, rather than along the Great Chain of Being.

Marsupial embryos and fossils like *Diarthrognathus* are what Carol Cleland called a "smoking gun" in her analysis of historical sciences. A smoking gun "is a trace(s) that unambiguously discriminates one hypothesis from a set of currently available hypotheses as providing 'the best explanation' of the traces thus far observed."[76] A "smoking gun" quite simply is pivotal evidence. Finding a broken bedroom window with pieces of glass strewn about inside the room may have many putative causes all of which could be framed as alternative hypotheses.

---

[72]Darwin, 1859, ibid., p. 84.

[73]Crompton, A.W. 1958. The cranial morphology of a new ictidosaurian. Proceedings of the Zoological Society of London, 130: 183–216.

[74]Starck, D. 1979. Vergleichende Anatomie der Wirbeltiere auf evolutionsbiologischer Grundlage. Springer, Berlin, p. 341.

[75]K.K. Smith, 2006. Craniofacial development in marsupial mammals: developmental origins of evolutionary change. Developmental Dynamics, 235, p. 1185. For more detail see Sánchez-Villagra, M.R., S. Gemballa, S. Nummela, K.K. Smith, and W. Maier. 2001. Ontogenetic and phylogenetic transformations of the ear ossicles in marsupial mammals. Journal of Morphology, 251: 219–238.

[76]Cleland, 2002, ibid., p. 481.

Finding the neighbor's boy baseball inside the bedroom resolves all rational disputes about the possible causes of the broken window. Similarly, *Diarthrognathus* and marsupial embryos resolve all rational disputes about the evolutionary origin of mammals.

Biology is not physics, and biological laws are not physical laws. Snakes are tetrapods that have lost the limbs that were ancestrally present. It is only natural that Darwin would note many exceptions to von Baer's laws (again citing Agassiz in this context). For example, the frog tadpole is a developmental stage, which frogs share with no other vertebrate animal. It is a special developmental stage that evolved in frogs in adaptation to the environment in which (most) frog species reproduce (and those frog species that do not so reproduce may lose the tadpole stage): frog eggs are deposited in ponds and slow-running bodies of water, which, when the tadpoles hatch, are rich in plant material that can provide a source of food for tadpoles with their specialized jaws and teeth, but not for adult frogs.[77] Hence Darwin's conclusion: "all organic forms have been formed on two great laws – Unity of Type, and the Conditions of Existence."[78] However, even if they afford many exceptions, von Baer's laws still capture important regularity in nature. Darwin published his final conclusion in the sixth edition of the "*Origin*" in 1866: "...community in embryonic structure reveals community of descent; but dissimilarity in embryonic development does not prove discontinuity of descent."[79]

Darwin's "long argument" published in 1859 is a book that changed the world. It is like a string of beads, every one of which a glowing "smoking gun" rendering the inference to evolutionary explanation a necessity rather than a mere possibility. At the present time, evolutionary theory has itself evolved into a highly complex, multifaceted body of thought. Subdisciplines are sometimes seemingly contradictory; some require special training for a proper understanding. And yet, evolutionary theory has become a predominant worldview. Evolutionary theory primarily addresses the problem of the origin of the diversity of organic beings, the diversity of plant and animal species that we observe today. In more recent times, however, evolutionary theory has gained currency far beyond its original confines. Attempts to understand the origin of the Earth, indeed of the universe, are now cast in an evolutionary context. And so are attempts to understand the origin and historical development of human culture and civilization, the origin of the powers of human cognition, and even the origin of moral values and ethical standards guiding and constraining everyday life in human society. Engineering uses computer software to simulate evolutionary processes such as (natural) selection in the attempt to optimize the design of complex mechanical systems such as aircraft. Simulations of

---

[77]Wassersug, R.J. 1975 The adaptive significance of the tadpole stage with comments on the maintenance of complex life cycles in anurans. American Zoologist, 15: 405–417.

[78]Darwin, 1859, ibid., p. 206.

[79]Peckham, M. 1959. The Origin of Species by Charles Darwin. A Variorum Text. University of Pennsylvania Press, Philadelphia, p. 703.

evolutionary processes are also used in the development of vaccines. Karl Popper, at this point a very familiar figure, once stated that "all life is problem solving."[80] One of the most important biologists of all times, Charles Darwin has shown us how nature goes about to solve problems. Today, humankind starts to see how a scientific understanding of those natural mechanisms can help to solve its own problems.

[80]Popper, K.R. 2001. All Life is Problem Solving. Routledge, London.

Index

Breinigsville, PA USA
06 March 2011
257024BV00008B/36/P

DATE DUE